Sustainability in Ruminant Livestock

Pramod Kumar Rout •
Basanta Kumara Behera

Sustainability in Ruminant Livestock

Management and Marketing

Springer

Pramod Kumar Rout
ICAR-CIRG
Mathura, Uttar Pradesh, India

Basanta Kumara Behera
Advanced Centre for Biotechnology
Maharshi Dayanand University
Rohtak, Haryana, India

ISBN 978-981-33-4342-9 ISBN 978-981-33-4343-6 (eBook)
https://doi.org/10.1007/978-981-33-4343-6

This Springer imprint is published by the registered company Springer Nature Singapore Pte Ltd.
The registered company address is: 152 Beach Road, #21-01/04 Gateway East, Singapore 189721, Singapore

Preface

The world food economy mainly depends on the shift in diet and food consumption patterns on the basis of livestock availability on quality and quantity basis. Consumption of animal proteins in the developing country has been growing over 4% per annum. Globally, livestock production is the largest user of agricultural land.

It provides livelihoods to about 1.3 billion people and contributes about 40% to global agricultural output. Its share of agricultural GDP is already 33% and is quickly increasing. Global meat production is projected to more than double from 229 million tonnes in 1999/2001 to 465 million tonnes in 2050, while milk output is set to climb from 580 to 1043 million tonnes. The prices of meats, milk, and cereals are likely to increase in the coming decades, dramatically reversing past trends. At present, the livestock sector is responsible for about 15% of global greenhouse gas (GHG) emissions, whereas it uses about 70% of the available agricultural land and represents about 8% of global water withdrawals. It generates 65% of human-related nitrous oxide, which has 296 times the global warming potential (GWP) of CO_2. Most of this comes from manure.

Keeping these facts in view, the proposed book begins with the modern concept on the domestication of ruminant animals farming system and their acceptability and demand. In this connection, their production system, integrated farming system, precision livestock farming, and sustainable livestock farming have been narrated with suitable and impressive model for the quick understanding of readers.

The second chapter explains the concept of implementation of mass balance approach to develop a perfect eco-friendly livestock farming independent of external energy input. In this connection, modern integrated farming system has been highlighted with much technical information explaining the design and evaluation of manure management systems and modeling and operational tools. The applied concept on mass balance operating process is explained on the basis of the difference between imported and exported mass across the farm boundary. Estimating mass balance can provide critical information for (comprehensive) nutrient management planning and to manage the movement of nutrients and manure. Estimation of whole-farm P mass balance is explained with a suitable model system.

The third and fourth chapters are about the latest information on domestic ruminant animals like cattle, goats, sheep, and buffaloes and their respective production traits for small bigger ruminants, care and management, new methods of breeding, selection for improving growth and milk production, genetic defects, housing system development, feeding requirement, and healthcare.

The subsequent chapters contain updated information on management aspects of livestock-farming strategies and generation of multiple-job opportunity while discussing various aspects of livestock farming operational protocols like housing and management; nurture of ram, ewe, lamb, and newborn calf and heifers; care of buck, DOE, and kid-nutrition flushing; concept zero grazing-system; diseases control and management; integrated farming for goats; crop–livestock integration, etc.

The most interesting and impressive aspects of this piece of work are crop–livestock integration; energy autonomy in cattle farming; value-added biopharmaceuticals from cattle farming; CAPEX for cattle farming; and concept on cattle farming base industrial complex. In addition, the authors have explained how industrial and intensive mixed farming systems can be developed to economize the livestock farming practice.

In addition, the subsequent chapters provide information on factors influencing livestock way of life and sustainable eco-friendly livestock farming. It includes climate change on livestock; detrimental effects of the industry; topographic and edaphic factors; thermal stress on livestock growth and development; socioeconomic development; and water requirements for livestock.

The book ends with the modern concept on supply chain management for livestock products and also explains how perishable livestock items can be well managed by cold logistic system with the help of modern information technology and packaging technology. In this connection, the authors have explained how inventory management and warehouse management will be helpful in supplying quality product to the end-user, timely. The last section of supply chain management highlights how quality control and quality assurance practices are helpful in risk-free deal of livestock products with maximum safe and secure.

Mathura, India Pramod Kumar Rout
Rohtak, India Basanta Kumara Behera

Acknowledgements

We are highly grateful to all those individuals who have encouraged us to take up the assignment. We are really indebted to Mrs. Asha Sharma, Katyayanee, Gopal and Mrs. Rasmirekha, Pradeep and Priya for their constructive remarks and encouragement during the preparation of manuscript. We are grateful to Prof. G.F.W Haenlein and Prof. K.C. Raghvan for their support in preparing this book. We are also thankful to Dr. Trilochan Mohapatra, Secretary, DARE and Director General, Indian Council of Agricultural Research, New Delhi for his kind support and encouragement. We are also thankful to Ms. Nandini and Mahima for their helping hands to shape the manuscript. The kind cooperation of our publisher Springer Nature is highly acknowledged.

About the Book

The global livestock sector is growing faster than any other agricultural subsector. It provides livelihoods to about 1.3 billion people and contributes about 40% to global agricultural output. Keeping these facts in view, for the first time the proposed book is designed to highlight the concept of integrated farming, precision livestock farming, and sustainable livestock farming on the basis of mass balance approach. This piece of work is restricted to domestic ruminants like cattle, goats, sheep, and buffaloes and explains the design and evaluation of manure management systems and modeling and operational tools. It also provides updated information on management aspects of livestock-farming strategies and generation of multiple-job opportunity while discussing various aspects of livestock farming operational protocols like housing and management; nurture of ram, ewe, lamb, and newborn calf and heifers; care of buck, DOE, and kid-nutrition flushing; concept zero grazing-system; disease control and management; integrated farming for goats; crop–livestock integration, etc. One of the new aspects of the book is supply chain management of domestic ruminants and their marketing process with special emphasis on inventory planning and control, impact of new technology development on supply chain, logistics operation of livestock products, and quality control and quality assurance of perishable livestock products.

Contents

About the Authors

Pramod Kumar Rout is presently working as principal scientist at CIRG, Mathura under Indian Council of Agricultural Research, New Delhi. He is a graduate in veterinary sciences from College of Veterinary Sciences and Animal Husbandry, Bhubaneswar and completed his master's and doctoral degree from Indian Veterinary Research Institute, Izatnagar, Bareilly. His field of specialization is genetics of disease resistance. He has also worked as postdoctoral fellow for 2 years at National Institute of Immunology, New Delhi. He has worked in Genetics and Genomics Division, Roslin Institute, Edinburgh, UK as DBT associateship. He is a researcher in the area of genetic improvement of livestock and improving selection decision with molecular genetics application. He also coordinates All India Coordinated Research Project on Goat Improvement. His research contributions are in the areas of genetic diversity of livestock, conservation of animal genetic resources and management, genetics of disease resistance, genetics of heat stress regulation in livestock, and milk genomics. He has also worked in the area of conservation of endangered goats in their natural habitat.

Basanta Kumara Behera was a professor of biotechnology at three distinguished Indian universities, where he had been regularly associated with teaching and research at postgraduate-level courses on the topic related to biomass process technology development, biopharmaceuticals, microbial process development, drug designing, bioenergy management, and biomass processing technology since 1978. In 2009, he joined an MNC company as adviser for specialty chemical production and drug design through microbial process technology. He is associated with national and international reputed companies as technical adviser for the production of biopharmaceuticals under cGMP norms. He has the credit of publishing authored books by CRC Press USA, SpringerNature Verlag, Germany, Elservier Inc., CABI, Nosworthy Way Wallingford Oxfordshire.

Abbreviations

ADG	Average daily gain
AEZs	Agroecological zones
AFL	Age at first lambing
AFS	Age at first service
AGPs	Antimicrobial growth promotants
AU	Animal unit
BCPs	Business continuity plans
BCS	Body condition score
BIFRA	Federal Institute for Risk Assessment
BLA	Biologics license application
BMECs	Bovine mammary epithelial cells
BW	Bodyweight
CAE	Caprine arthritis encephalitis
CAFOs	Concentrated animal feeding operations
CAGR	Compound annual growth rate
CEE	Common environmental effect
cGMP	Current good manufacturing practice
CH_4	Methane
CID	Dissociation constants (pK_a)
COGS	Cost of goods sold
CRM	Customer relationship management
DEBT	Decreasing Earning Before Time
DGC	Dairy Goat Co-operative
DMAIC	Define, measure, analyze, improve, and control
DMI	Dry matter intake
EAEMP	European Agency for the Evaluation of Medicinal Products
EC	Electrical conductivity
EGFR	Epidermal growth factor receptors
EOQ	Economic order quality
EPA	Environmental Protection Agency
ERP	Enterprise resource planning
EU	European Union
EWG	Environmental Working Group

FAO	Food and Agriculture Organization
FCE	Feed conversion efficiency
FDA	Food and Drug Administration
FEC	Feed energy content
FFDCA	Federal Food, Drug, and Cosmetic Act
FIFO	First in first out
FMD	Foot and mouth disease
FMEA	Failure mode and effects analysis
GCMMF	Gujarat Co-operative Milk Marketing Federation Ltd.
GFY	Greasy fleece yield
GHG	Greenhouse gas
GLEAM	Global Livestock Environmental Assessment Model
GLSC	Green logistics and supply chain
HA	Direct heritability
Hb	Hemoglobin
HRI	High-income countries
HSP/s	Heat shock protein/s
HSUS	Humane Society of the United States
ILRI	International Livestock Research Institute
INDA	Investigational New Drug Application
IOS	Interorganizational information systems
IPBES	Intergovernmental Science-Policy Platform on Biodiversity and Ecosystem Services
IPCC	Intergovernmental Panel on Climate Change
IRI	Intermediate income countries
ISO	International Standard Organization
IT	Information technology
JMPR	Joint Meeting on Pesticide Residues
JMPS	Joint Meeting on Pesticide Specifications
LAC	Latin America and the Caribbean
LCA	Life cycle assessment
LEAP	Livestock Environmental Assessment and Performance
LIFO	Last in first out
LLS	Livestock's Long Shadow
LPELC	Livestock and Poultry Environmental Learning Community
LRI	Low-income countries
mAbs	Monoclonal antibodies
MAP	*Mycobacterium avium* sub sp.
MRO	Maintenance, repair, and operating
MRP	Material requirements planning
MRPH	Manufacturing resource planning
mtDNA	Mitochondrial DNA
N	Nitrogen
N_2O	Nitrous oxide

Na	Sodium
NDA	New drug application
NO_3	Nitrate
NRDC	Natural Resources Defense Council
P	Phosphorus
PCV	Packed cell volume
PHSA	Public Health Service Act
PSA	Pressure swing adsorption
QA	Quality assurance
QAPs	Quaternary ammonia products
QC	Quality control
RBCs	Red blood cells
RFI	Residual feed intakes
RFID	Radio-frequency identification technology
ROS	Reactive oxygen species
RTVN	Real time value network
SAMMs	Safety assessment of marketed medicines
SCM	Supply chain management
SDGs	Sustainable Development Goals
SKU	Stock-keeping unit
SO_4	Sulfate
SOC	Soil organic carbon
SOD	Superoxide dismutase
THI	Temperature-humidity index
TNF	Tumor necrosis factors
TPSCM	Third-party supply chain management
TQM	Total quality management
UHT	Ultra-high temperature treatment
UNFCCC	United Nations Framework Convention on Climate Change
USA	United States of America
USDA	United States Department of Agriculture
USPTO	United States Patent and Trademark Office
WBC	White blood cell
WIP	Work-in-process
WMO	World Meteorological Organization
YBP	Years before the present

1 Prelusion Significance of Livestock

1.1 Etymology and Legal Definition

Livestock carries a broad meaning; it is a compound word as "live" and "stock" [1]. The word livestock was first used between 1650 and 1660. The disaster assistance legislation in 1988 defined the term as "cattle, sheep, goat, swine, poultry (egg-producing poultry), equine animals used for food, fish used for food. Livestock is usually defined to refer to horses, mares, mules, jacks, jennies, colts, cows, calves, yearlings, bulls, oxen, sheep, goats, lambs, kids, hogs, shoats, and pigs. States regulate livestock to promote health and welfare and prevent damage and injury caused by their uncontrolled roaming on others' property.

So, in general, we can define livestock as domesticated animals, mainly based on agricultural setup, and produce commodities such as meat, eggs, milk, fur, leather, and wool. Earlier the term "livestock" was referred as used as "'rare breed for consumption", other times it refers only to farmed ruminants, such cattle, and goats. In United States horses are also included in livestock. The USDA categorized pork, veal, beef, and lamb as livestock, and all livestock as red meat. Poultry and fish are excluded from livestock.

1.2 Domestication

1.2.1 History of Domestication

The domestication of plants and animals over the past 11,500 years has transformed the earth's biosphere and affecting evolution of other species including human beings. Animal domestication was proposed based on three separate pathways that animals followed in to a domesticated relationship with humans', i.e. commensal pathway, a prey pathway, and a directed pathway [2]. The genetic differentiation of domestic and wild population can be framed within two major events. The first event, the domestication traits that developed in captive animals and improved traits

P. K. Rout, B. K. Behera, *Sustainability in Ruminant Livestock*,
https://doi.org/10.1007/978-981-33-4343-6_1

that appeared distinct in wild and domestic populations [3]. Animal domestication has been proposed from different angles; however, the challenges remain to identify the genes that were responsible for differentiating domestic populations from their wild ancestors. It has been suggested that crop cultivation have begun independently in 20 different regions [4]. However the early animal domestication had evolved in three regions, i.e. the Near East, Central China, and the Andes. The animal domestication started in the region near to agricultural societies and near to the regions of plant domestication. The animal domestication took place within the framework of three pathways, i.e. commensal pathway, a prey pathway, and a directed pathway [2].

Commensal Pathway

Dogs were domesticated by commensal pathway across Eurasia well before the domestication of other animals [5]. Cats also followed the commensal pathway in both Europe and subsequently transported to Cyprus by 11,000 years BP [6]. Pigeon were attracted to human habitat and pigeon-keeping was first observed in Mesopotamia and Egypt [7]. Guinea pigs and rats also followed commensal pathway. Pigs must have followed commensal pathway as they consume human waste and convert it to productive protein. Pigs were associated with millet cultivators in Central China [8], whereas in Western Eurasia pigs were associated with villages of the oak woodland zone of the northern Fertile Crescent [9]. Several other species followed commensal pathway in eastern Asia in rice cultivation area. The chickens were domesticated via commensal pathway and they were domesticated as early as 8000 years BP [10]. However it was again suggested that chickens must have been domesticated by at least 4500–4000 years BP [11]. Ducks and goose domestication in East Asia near paddy field, carp were associated with rice agriculture. Similarly fresh water fishes were associated with rice cultivation near Yangtze River [12].

Prey Pathway

The meat animals are generally attracted by prey pathway. Sheep, goats, and cattle were domesticated via prey pathway in western Eurasia between 10,500 and 10,000 years BP [9]. Cereals and pulses were undergoing domestication in the same region during the same period [13]. Zebu cattle were independently domesticated through the prey pathway in the Indus Valley, where they are associated with early cultivating villages between 9000 and 7000 years BP [11]. Similarly, river and swamp species of water buffalo were independently domesticated following the prey pathway. The river buffalo was domesticated in the lower Indus Valley region and western India by 4500–4000 years BP [11]. Swamp buffalo was likely domesticated between eastern India and peninsular Southeast Asia before 3000–2500 years BP. Similarly sheep and goats followed prey pathway in Western Eurasia region in 8000–7500 years BP [14].

Directed Pathway

Donkeys were domesticated in southern Fertile Crescent region. Bactrian camels in the cool deserts of Central Asia [15] and dromedary camels were domesticated in the deserts of Arabia [16]. Horses were domesticated in the temperate grasslands of Central Asia and horses were first tamed as early as 7000 years BP. Honey bees were first associated with Jordan Valley [17]. The global patterns of animal domestication suggest that first domestication of animals in different regions followed commensal or prey pathway.

1.2.2 Domestication of Major Livestock Species

The people across the world lived as hunter–gatherers and were facing climate vagaries. A major change occurred in human culture and people started to settle at one place by adopting agriculture and domesticating animals. This phenomenon, known as the Neolithic Revolution, happened independently in different parts of the world and at subsequent times between 11,000 and 4500 years before the present (YBP). In particular, two small areas of Eurasia, the Fertile Crescent and Southeast Asia, were the earliest centres of domestication for a remarkable array of agricultural crops and farm animals: wheat, rye, barley, lentils, goats, sheep, cattle, pigs, and buffalo [18–20]. Among these, goats were the first livestock species to be domesticated in the Fertile Crescent [21, 22]. As pointed out by Zeder [23], domestication is a gradual process, not an instantaneous event. The dog was domesticated in Europe and the Far East from about 15,000 years ago. Goats and sheep were domesticated in multiple events sometime between 11,000 and 5000 years ago in Southwest Asia. Pigs were domesticated by 8500 BC in the Near East and 6000 BC in China. Domestication of the horse dates to around 4000 BC. Cattle have been domesticated since approximately 10,500 years ago. Chickens and other poultry may have been domesticated around 7000 BC.

1.2.2.1 Cattle

There are two major types of domestic cattle are found such as zebu (*Bos indicus*) and *Bos taurus*. *Bos indicus* has prominent thoracic hump and *Bos taurus* has no hump [24, 25]. However, these two species fully interbreed, and a meta-analysis of different microsatellite datasets revealed taurine-zebu admixture over Europe, Southwest Asia, and Africa [24–27]. Molecular evidence suggests that these two species came from two independent domestication events: zebu cattle were domesticated in the Indus Valley region ca. 8000–7500 BP, whereas taurine cattle were domesticated in Anatolia 10,500–10,000 BP [25, 27–30]. However, Larson and Burger [31] suggested that only the latter was domesticated, while zebu may have resulted from the introgression of wild zebu populations into taurine cattle that were transported eastward. The first cattle herd book was published in Britain in 1822. Since that time, stronger selection pressures have been applied to local populations followed by standardization of the desired conformation and performance, such as high milk yield for dairy cattle breeding programmes [24]. This led to an isolation of breeds from each other and more than 1000 breeds are recognized all over the world, which could have caused a genetic drift and inbreeding and perhaps a fitness decrease [25, 27, 32]. Nevertheless, gene flow between neighbouring regions did not completely stop, as deliberate upgrading was realized in order to increase production.

1.2.2.2 Sheep

The wild ancestors of the domestic sheep are probably the mouflon (*Ovis musimon*) and the urial (*Ovis orientalis*) [24, 25, 29]. Both archaeological and genetic data spot the domestication centre of sheep in eastern Anatolia and North-West Iran [25] between 8500 [24] and 12,000 years ago [27]. The sheep mitochondrial DNA polymorphism diversity and single-nucleotide polymorphism (SNP) diversity seem to support an absence of a genetic bottleneck, and thus domestication occurred from a broad genetic base [25, 33]. Sheep were first farmed for access to meat before human-mediated specialization for wool and milk commenced ca. 4000–5000 years ago [33]. It has recently been shown that particular regions of the genome contain strong evidence for accelerated change in response to artificial selection, such as the removal of horns, likely to be one of the oldest morphological modifications that accompanied domestication and a trait now common across many modern breeds [33].

1.2.2.3 Goat

The wild ancestor of goat is the bezoar, *Capra aegagrus* [24, 27, 29]. The first archaeological evidence of goat domestication traces back in the Fertile Crescent about 10,000 years ago [24, 29, 34]. A large-scale analysis of current bezoar mitochondrial DNA (mtDNA) polymorphism over its whole geographic distribution suggested that the domestication process occurred over a very large area encompassing eastern Anatolia and North-West Iran [25]. Additional primary centres of goat domestication, including the Indus Valley, Southern Levant, and China, have not been convincingly demonstrated yet [34]. Analysis of the goat mitochondrial DNA polymorphism of the main haplogroup (representing more than 90% of the haplotypes) strongly supports the absence of bottleneck at the domestication time in goats [25]. Besides, goat mtDNA polymorphism also suggests high historical gene flow among continents, which already occurred during the Neolithic expansion into Europe [24]. The extraordinary adaptability and hardiness of goats favoured their rapid spread over the Old World [34]. Goats have successfully adapted to desert, mountainous, and tropical areas where other livestock species would not thrive [34]. Between the fifteenth and eighteenth centuries, goats were transported to America and Oceania [34]. Over the course of domestication, several morphological traits were modified, such as horn and ear shapes, the presence of wattles, long hair, and coat colours, which were driven probably by intentional selection as well as by genetic drift, isolation, and founder effects [34]. Throughout the ages, goats have been raised for milk production and cheese, meat, and skin and fibre commodities such as leather, mohair wool, and cashmere hair [34]. Breeds also show strong differences in their physiological capacity of adaptation to extreme conditions of temperature and humidity and differ in feed efficiency, behaviour, and resistance to infectious and parasitic diseases [34].

1.2.2.4 Horse

The wild ancestor of domestic horse is the now extinct *Equus ferus* from Central Asia [28]. The Asian wild horse, *Equus przewalskii* [29], also significantly contributed to the genetic makeup of domestic horses [35]. Even though there have been no confirmed sightings of wild Przewalski's horses since 1966, the species has been maintained in captivity for the last 90 years [29]. In addition to Przewalski's horse, a third divergent lineage corresponding to a wild population that inhabited the Holarctic region has also contributed to the genome of modern domestic horses [35]. Both archaeological and genetic evidence strongly support the onset of domestication of horse in the western Eurasian Steppes of Ukraine dating to 5500 years ago [29, 35]. Domestic horses exhibit remarkable variation in coat colouration, including the bay or bay-dun wild-type phenotypes, other basic colours like chestnut and black, as well as dilution (e.g. cream and silver), and spotting patterns (e.g. leopard complex, tobiano, and sabino) [35]. Horse locomotion has also been recurrently selected, including their ability to perform alternate gaits, such as four-beat, lateral, or diagonal ambling [35].

1.3 Public Acceptability and Demand for Livestock Products

Population growth, urbanization, and rise in per capita income have strongly influenced the livestock product consumption in different parts of the world. Urbanization has considerable impact on demand for livestock products. Urbanization brings improvements in infrastructure including storage of perishable goods and make livestock products to be traded more widely [36]. The increased demand for livestock products is due to income growth. As income grows, so does expenditure on livestock products increases [37]. Food demand for livestock products will nearly double in sub-Saharan Africa and South Asia, from some 200 kcal per person per day in the year 2000 to around 400 kcal per person per day in 2050.

Global livestock production has increased substantially since the 1960s. Beef production has increased two fold while over the same time chicken meat production has increased by a factor of nearly 10, this increase is due to increases in both number of animals and productivity. Carcass weight increased by about 30 percent for both chicken and beef cattle from the early 1960s to the mid-2000s, and by about 20 percent for pigs [38]. Carcass weight increases per head for camels and sheep are much less, about 5 percent only over this time period. Increases in milk production per animal have amounted to about 30 percent for cows' milk over the same time period [38]. These changes have been accompanied by substantial shifts in the area of arable land, pastures, and forest. Arable and pasture lands have expanded considerably since the early 1960s, although the rates of change have started to slow [36]. Over the last 20 years, large forest conversions have occurred in the Amazon Basin, Southeast Asia, and Central and West Africa, while forest area has increased owing to agricultural land abandonment in the Eurasian boreal forest and parts of Asia, North America, and Latin America and the Caribbean (LAC) [39].

The domesticated livestock population size during 2018 with respect to their use is presented in Table 1.1. Similarly the production of meat and milk from domesticated livestock species in major countries is presented in Table 1.2. Global Population growth trend and growth rate of domesticated livestock species during 1967–2017 are presented in Fig. 1.1a, b. All the livestock species are showing increasing trend during the last five decades and the population growth rates are 10.83%, 13.28%, 2.87%, and 8.40% for bovine, goat, sheep, and pig, respectively. The carcass weight and milk production of bovine showed an increase of 16.9% and 49.6% during 1967–2017 (Fig. 1.2). Similarly, the carcass weight of goat and sheep showed an increase of 16.67% and 12.08%, respectively (Figs. 1.3 and 1.4), however, there was no change in milk production per animal during last 50 years (Figs. 1.3 and 1.4). The pig showed an increase of 25.3% in carcass weight during 1967–2017 globally (Fig. 1.5).

It has been estimated that the livestock production growth will be from confined systems by 2030; however, Africa and Asia will witness much slower growth from this system. Similarly, the crop production growth will come mostly from yield increases rather than from area expansion; the increases in livestock production will come about more as a result of expansion in livestock numbers in developing countries, particularly ruminants. In developed countries, carcass weight growth will contribute an increasing share of livestock production growth as expansion of numbers may contract in some regions.

In the intensive mixed systems, food-feed crops are vital ruminant livestock feed resources. The prices of food-feed crops are likely to increase at faster rates than the prices of livestock products [40]. The prices of meats, milk, and cereals are likely to increase in the coming decades. Bioenergy demand is projected to compete with land and water resources, and this will exacerbate competition for land from increasing demands for feed resources. The production of alternative feeds for ruminants in the more intensive mixed systems, however, may be constrained by both land and water availability, particularly in the irrigated systems [41]. Growing scarcities of water and land will require substantially increased resource use efficiencies in livestock production to avoid adverse impacts on food security and human well-being goals. Higher prices can benefit surplus agricultural producers, but can reduce access to food by a larger number of poor consumers, including farmers who do not produce a net surplus for the market. As a result, progress in reducing malnutrition is projected to be slow [40]. Livestock system evolution in the coming decades is inevitably going to make balance between food security, poverty, equity, environmental sustainability, and economic development.

1.4 Nutrient for Biological Process

The nutritional needs of farm animals with respect to energy, protein, minerals, and vitamins have been well studied. The requirement of nutrient for various physiological stages has been determined for ruminants and non-ruminants, which were designed to assess the nutritional and productive consequences of different feeds

Table 1.1 Major domesticated livestock species in relation to their number and uses

Livestock species	Number of population (in millions)	Leading countries with population size (in millions)		Primary use production
Ruminants				
Cattle	1489.744	Brazil	213.523	Meat, milk, hides
		India	184.464	
		China	63.417	
Sheep	1209.467	China	164.079	Wool, meat, hides
		Australia	70.067	
		India	61.666	
Goats	1045.915	China	138.383	Milk, meat, hair, hides
		India	132.749	
		Sudan	31.837	
Buffaloes	206.600	India	114.151	Milk, meat, hides
		China	27.118	
		Brazil	1.390	
Camels	35.525	Sudan	4.872	Transport, milk, meat, draft
		China	0.323	
		India	0.311	
Non-ruminants				
Chicken	23,707	China	5372.559	Meat, eggs, Feathers
		Indonesia	2384.147	
		United States	1973.38	
		Brazil	1468.352	
		India	801.09	
Pigs	978.332	China	447.175	Meat
		United States	74.550	
		Brazil	41.443	
		Indonesia	8.542	
		India	8.485	
Turkeys	466.787	United States	244.75	Meat, eggs, feathers
		Brazil	33.781	
		China	0.083	
Ducks	1124.90	China	692.691	Meat, eggs, feathers
		Indonesia	60.011	
		India	19.64	
		United States	7.373	
Horses	57.780	United States	10.479	Draft, riding, sport, occasionally meat
		Brazil	5.751	
		China	3.438	

(continued)

Table 1.1 (continued)

Livestock species	Number of population (in millions)	Leading countries with population size (in millions)		Primary use production
		India	0.625	
		Indonesia	0.421	
Donkeys	50.453	China	2.677	Draft, transport
		Brazil	0.822	
		India	0.229	
		United States	0.051	
Mules	8.525	Brazil	1.252	Draft, transport
		China	0.811	
		India	0.230	

www.fao.org (Data accessed from FAO on 24/08/2020)

for the animal once intake was known. The dynamics of digestion, feed intake, and animal performance had to be predicted in many livestock species with high accuracy. It is necessary to determine the prediction of animal growth, body composition, feed requirements, and the outputs of waste products from the animal and production costs. This aspect will help to improve the efficiency of livestock production and fulfilling the consumers' requirement and the demands of regulatory authorities. It is necessary to apply genomics, transcriptomics, proteomics, and metabolomics to the field of animal nutrition and predictions relating to growth and development. Simultaneously it is also required to analyse nutritional requirement as well as various nutritional stress for the ruminants in pastoral and extensive mixed systems in developing countries [42].

Poor nutrition is one of the major production constraints in smallholder systems, particularly in Africa and in Asia. It has been tried by researchers to improve the quality and availability of feed resources, including sown forages, forage conservation, the use of multi-purpose trees, fibrous crop residues, and strategic supplementation. There are also prospects for using novel feeds from various sources to provide alternative sources of protein and energy, such as plantation crops and various industrial by-products. The prevalence of mixed crop–livestock systems in many parts of the world, closer integration of crops and livestock in such systems can give rise to increased productivity and increased soil fertility [43]. In such systems, smallholders use crops for multiple purposes such as for their own food and livestock feed; therefore, crop breeding programmes may target to fulfil the requirement of smallholder production system.

There is considerable work being carried out to address the issues associated with various antinutritional factors. These include methods to reduce the tannin content of tree and shrub material. The use of essential oils that may be beneficial in ruminant nutrition and that can lead to beneficial effects on livestock performance. Enzymes are widely used to achieve substantial gains in feed conversion efficiency. If

Table 1.2 Production of animal products

Animal species	Animal product	Countries	Production (Tonnes)
Cattle	Milk	United States of America	98,690,477
		India	89,833,590
		Brazil	33,839,864
		Germany	33,064,833
		China	31,165,090
	Meat	United States of America	12,219,203
		Brazil	9,900,000
		China	5,810,609
		Mexico	1,980,846
		India	947,873
Sheep	Milk	Turkey	1,446,271
		China	1,180,276
		Greece	753,819
		Syrian Arab Republic	647,311
		Romania	626,145
	Meat	China	2,422,857
		Australia	735,009
		Sudan	264,000
		India	229,834
		United States of America	71,758
Goat	Milk	India	6,098,730
		Sudan	1,151,000
		Bangladesh	1,122,646
		Pakistan	915,000
		Mali	526,130
	Meat	India	6,098,730
		China	2,329,767
		Mexico	39,851
		Brazil	36,546
		United States of America	9169
Buffalo	Milk	India	91,817,140
		Pakistan	28,109,000
		China	3,003,323
		Egypt	2,120,365
		Nepal	1,338,277
	Meat	India	1,662,383
		Pakistan	949,000
		China	644,529
		Italy	22,328
		Bangladesh	6829
Camel	Milk	Somalia	958,079
		Kenya	854,669
		Mali	603,821

(continued)

Table 1.2 (continued)

Animal species	Animal product	Countries	Production (Tonnes)
		Ethiopia	176,113
		Niger	105,967
	Meat	Sudan	145,000
		Kenya	72,780
		Somalia	46,663
		Egypt	36,398
		China	17,055
Pig	Meat	China	54,983,905
		United States of America	11,942,965
		Brazil	3,787,660
		Mexico	1,502,521
		India	296,145

www.fao.org. Data accessed from FAO on 24/08/2020

intensification of production will not be possible, then there are many ways in which nutritional constraints could be addressed. One area of high priority for additional exploration, which could potentially have broad implications for tropical ruminant nutrition, is microbial genomics of the rumen, building on current research into the breaking down of lignocellulose for biofuels [44].

Livestock productivity can be improved in semiarid and arid areas by providing the most information on drought prediction, so that herders can better manage the complex interactions between herd size, feed availability, and rainfall [44]. The future of livestock nutrition depends on increased feed conversion efficiency to keep livestock production profitable. Public health issues will become increasingly important, such as concerns associated with the use of antibiotics in animal production, including microbiological hazards and residues in food [45]. The World Health Organization recommended that all sub-therapeutic medical antibiotic use to be stopped in livestock production in 1997. Similarly, the sub-therapeutic treatments such as growth promotants should not be used in livestock production. All antibiotics as growth promoters were banned in the European Union (EU) in 2006, but not all countries have made the same choice as the EU. Similarly, certain hormones can increase feed conversion efficiencies, particularly in cattle and pigs, and these are used in many parts of the world. The EU has also banned the use of hormones in livestock production.

The globalization of the food supply chain will continue to raise consumer concerns for food safety and quality. The livestock nutrition will be more dependent on mitigation of greenhouse gas emissions. Improved feeding practices (such as increased amounts of concentrates or improved pasture quality) can reduce methane emissions per kilogram of feed intake. Many specific agents and dietary additives have been proposed to reduce methane emissions, including certain antibiotics, compounds that inhibit methanogenic bacteria, probiotics such as yeast culture,

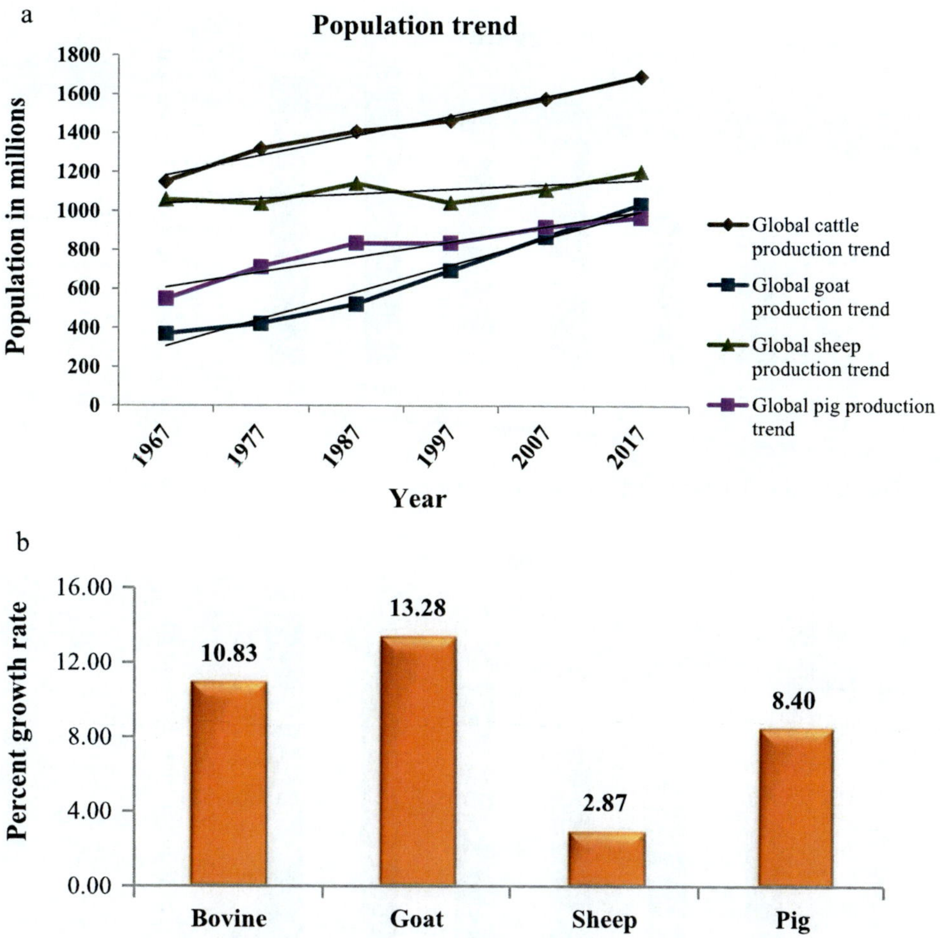

Fig. 1.1 The population growth trend of domesticated livestock species during 1967–2017 (**a**) and population growth rate (**b**)

and propionate precursors such as fumarate or malate that can reduce methane formation [46].

1.5 Production System

Livestock production and managemental practices vary around the world. Therefore different agro-ecological zones are defined based on climate, landforms, soil, land cover, and land use. The agro-ecological zones are broadly classified into five categories including tropics, subtropics, temperate, boreal, and artic [47]. Agro-ecological zones (AEZs) have been established to assess agricultural resources by FAO [48].

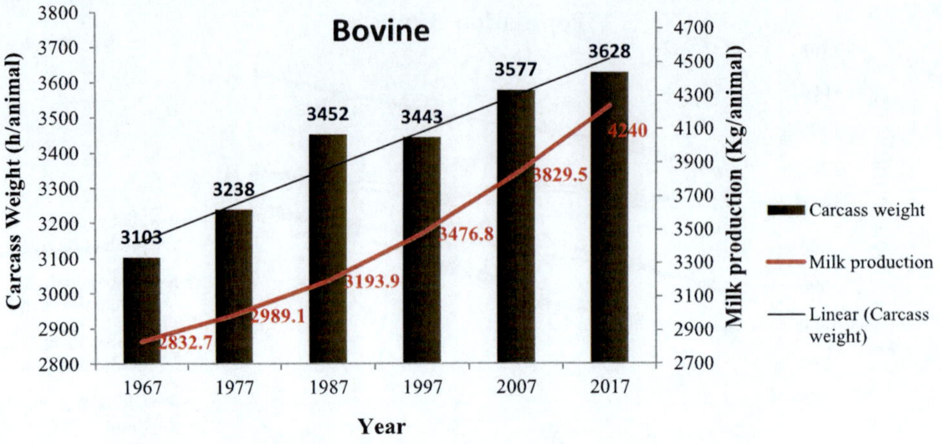

Fig. 1.2 Global bovine milk production and carcass weight trend during 1967–2017

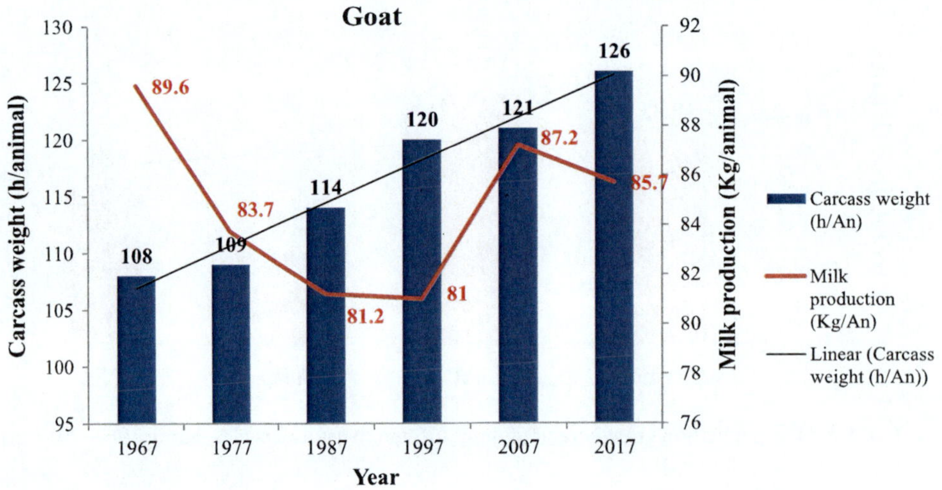

Fig. 1.3 Global milk production and carcass weight trend in goat during 1967–2017

Global livestock production system has been classified based on value of production and also based on types of rearing. FAO classified livestock production system mainly as two types: (1) where less than 10% of the total value of production comes from non-livestock farming practices and (2) mixed farming system where more than 10% of the total value of production is from non-livestock farming activities [49]. A grassland based production system and landless livestock production system are adopted in developed countries including Central and South America. Grazing and mixed farming systems are concentrated in temperate and tropical highland zones, humid and sub-humid tropical and subtropical zones, and arid and semiarid tropical and subtropical zones of the world. A landless livestock production system is otherwise known as intensive production system is characterized by less than 10%

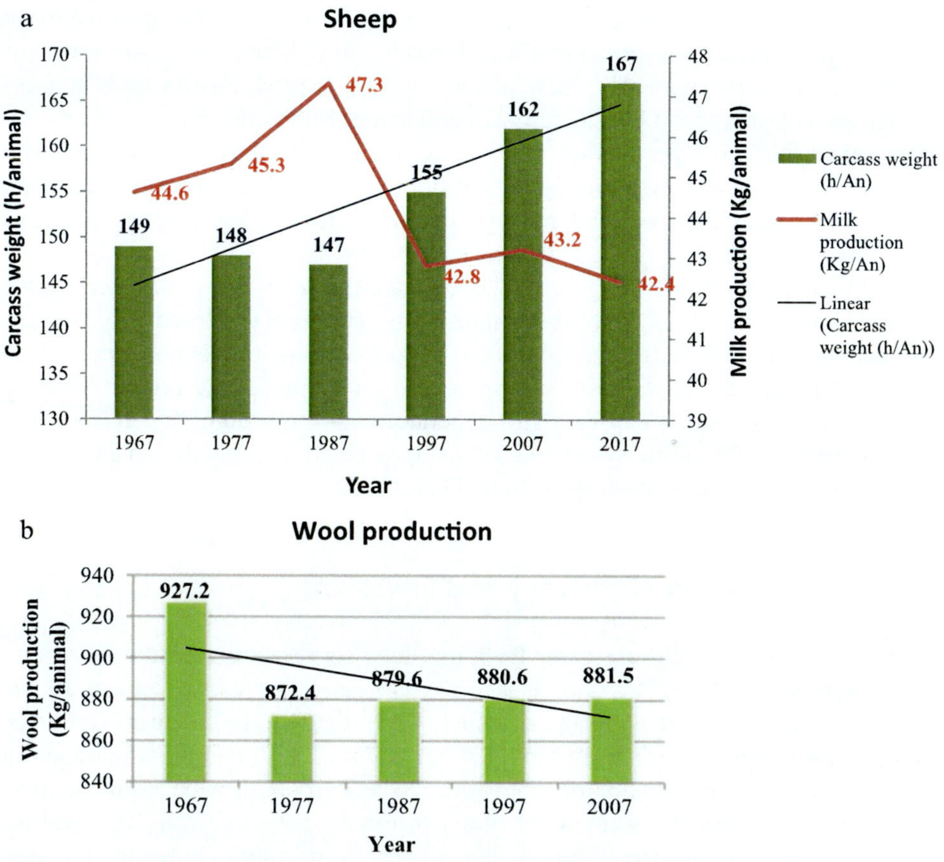

Fig. 1.4 Global milk production and carcass weight (**a**) and wool production (**b**) trend during 1967–2017 in sheep

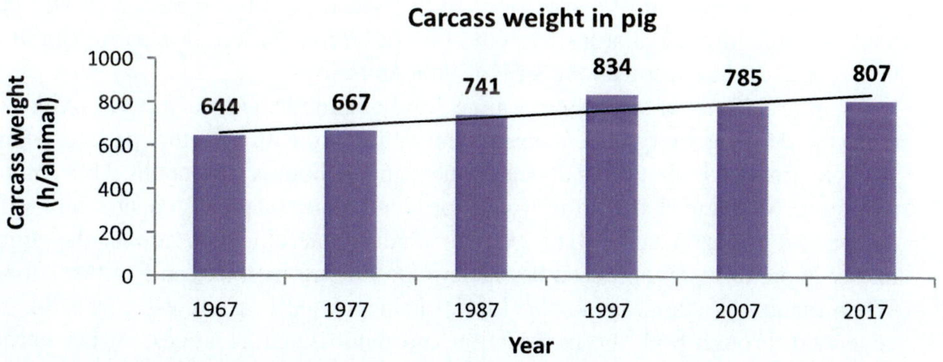

Fig. 1.5 Global carcass weight production trend during 1967–2017 in pig

of animal feed is produced on farm and animal density is more than 10 livestock units per hectare. A grassland based production system (extensive system) is characterized by more than 10% of animal feed is produced on farm and animal density is less than 10 livestock units per hectare. Mixed farming system represents livestock and crop production system and it includes rain-fed as well as irrigated mixed farming system. Globally, mixed farming systems provide more than 54% of total meat and 92% of total milk production followed by landless systems and grazing systems [50].

In India, Mixed farming system dominates the livestock production system in both arid/semiarid and humid/sub-humid climatic zones. Mixed irrigated system supports the production system and supports crop production more feasible [47]. Mixed rain-fed farming system is also wide spread and region is especially difficult for livestock due to high temperature and high humidity. In rain-fed and irrigated mixed farming system also wide-spread and integrated crop/animal system include rice/wheat/goat/sheep in India [51].

1.5.1 Production Efficiency and Emissions

The important greenhouse gases from livestock sources are methane, nitrogen, and nitrous oxide. Therefore the livestock sector has to follow strategy to reduce environmental burden and simultaneously fulfill the food requirement for the growing world population. The environment emission on different production system is composed as carbon dioxide equivalent. The carbon dioxide equivalent is the standard unit used to account for global warming potential [52]. The greenhouse gas emission in livestock production comes from enteric fermentation, manure storage, and feed production. Enteric fermentation occurs due to break down of biomass in rumen, due to action of bacteria, protozoa, and fungi. Ruminants convert cellulose and semi-cellulose into energy need of individuals. Plant biomass in the rumen is converted into volatile fatty acids, which pass the rumen wall and go to liver through the circulatory system. Enteric fermentation producing about 2.8 Gigatonnes and produce about 39% of total emission.

Manure acts as an emission source for both methane and nitrous oxide. The emission depends on type of management and composition of the manure. Manure storage produced about 0.71 Gigatonnes and accounts for about 10% of total emission. Methane and nitrous oxide production in manure depends on organic matter and nitrogen content of excreta. Under anaerobic conditions, the organic matter is partially decomposed by bacteria producing methane and carbon dioxide. When manure is stored as a solid (dung) or deposited in pastures, nitrous oxide is generated through both the nitrification and denitrification process of the nitrogen contained in manure.

Methane is formed by methanogenic archaea in rumen. Methane is mainly formed in rumen (90%) and about 10% of methane is formed in hind gut. Methane production depends on various factors such as composition of diet and level of feed intake in the ruminant. Ruminant production system with higher intake of fibrous feed may be associated with high levels of GHG emission per unit products. Ruminant consumes human inedible feed (fibre rich forages and by-products) and produce CH_4 in this process. The consumption of fibrous feeds is associated with high levels of GHG emissions per unit product. The enteric CH_4 emission can be reduced by different dietary options [53–55]. The addition of dietary fat or oil reduces CH_4 production by about 5% with 1% addition of lipid to the diet [56].

Dietary lipid supplementation decreases CH_4 production by reducing the activity of methanogens and protozoan numbers and reducing fermentation of substrate in the rumen. Monensin, rumen modifiers, reduces CH_4 production but the use of monensin in livestock industry is not allowed. Yeast culture is used to maintain the milk yield and production efficiency, thus may decrease CH_4 production. Nitrate is also highly effective in reducing CH_4 in ruminant, but the risk of nitrate poisoning is important [57]. Besides this number of dietary additives has shown promising effect on CH_4 production in-vitro and not showed effectiveness in vivo. As the production system becomes efficient, the waste excretion decreases per unit product. The N and CH_4 efficiency are more observed at low level of production. The ability of ruminants to convert human inedible products in to human-edible products is an important concern with respect to global food security. The ruminant production is mainly dependent on natural resources and nutrients from various by -products. The ruminants are ultimately used as resources for animal proteins, and plays a crucial role in food security. Intensive production system has become source of N pollution and GHG emission.

1.6 Integrated Farming System

The mixed crop–livestock system is the major production system and is more sustainable as the resource used effectively and soil fertility remains intact. Mixed livestock farming is a self-supportive ecosystem which was based on symbiotic relationship with different components. The system provides energy security, soil and water conservation, better utilization of space, increasing cropping intensity, and manages pest incidence. This makes the farming system more resilient and highly productive. Integrated farming system is the best way of sustainable resource management. The integration is approached with crop, pond, horticulture, and animals. The animals such as cow, goat, rabbits, pig, and chicken can be introduced to provide waste products as source of nutrients and other functional inputs. The system provides effective linkages between various components, thereby reducing cost of production, effort, risk and enhances diversity and nutrition.

Livestock–fish integration maximizes food production and economic return per unit area of land. Its main benefit comes from the conversion of livestock wastes (faeces, urine, stall/pen washing, spilled animal feed into protein). The livestock wastes entering the pond, thus, enhancing microbial and plankton production which contributed to fish nutrition. Integrating goat farming with cropping systems like agro forestry models such as agri-silviculture, hortipasture, boundary plantations, wind belts, and silvipasture models with trees and grasses can be established in wastelands to supplement farm income.

1.7 Precision Livestock Farming

It is required to increase biological and production efficiency of livestock farming to increase profitability. Precision livestock farming is an integrated livestock production system based on measurement, management, and marketing of animals according to their individual merit. Generally, livestock production and management address the group management and care, although significant individual variation in performance observed between animals. It was really difficult to identify and analyse the performance of an individual in a large commercial unit or smallholder production system. Precision livestock farming uses information technology for assessing the individual animal with respect to physical resource and brings improvement accordingly to optimize economical farming. This helps us to identify poor performance animals in the flock and cull those individuals from flock. The electronic identification has made it possible to identify individual in accurate manner and faster rate. Profit is the major driver of livestock production system. Profit will be determined by effective production and optimum management practices. Feed is one of the major inputs and varies from 58% to 70% for livestock farming input cost. However high input cost can be reduced by implementing precision grazing and feeding management regime. Similarly the labour cost constitutes 18–25% of farming input cost and can be reduced by effective management and use of technology.

1.8 Sustainable Livestock Farming

Livestock production system characterization is essential for sustainable production. Sustainable livestock production and agricultural production must be viewed as flow of energy among soil, plant, and livestock as a cycle. Sustainability in livestock production system needs to be analysed based on indicators such as ecological adaptability, viable economy, and energetic efficiency of the production system. Sustainability of intensive and extensive livestock production system can be analysed through tools such as life cycle assessment (LCA) and material and energy flow analysis. The environmental footprint of every production system needs to be analysed. Sustainable livestock management is closely associated with energy flow. Sustainable livestock farming can bring resource efficiency, strengthens resilience, and secures social responsibility of agriculture and food system to ensure nutrition as well as food security [58]. Sustainable key factors are water conservation, soil conservation, and soil nutrients management, energy recycling system, biodiversity conservation, and continuity of biodiversity.

The various beneficial aspects of livestock farming are development of global agricultural economy, rural livelihood up-gradation, rural women's empowerment, sustainable rangeland management, conservation of domestic animals, and enhancing soil fertility and nutrient cycling. As per the report of FAO [59] the grazing land is more than the cropping land. The grazing land covers about 60% of the total agriculture land and supports 360 million cattle and 600 million sheep and goats. About 10% of the world's beef meat and 30% sheep and goats meat are produced from grazing animals worldwide. In addition, sustainable livestock farming not only save the land for cropping but also helpful in maintaining nitrogen and phosphors balance. The goat's excreta are rich in nitrogen and phosphors. By adapting biogas production techniques from ruminant farm organic wastes, methane gas can be produced to meet the energy demand of farm house, and the slurry leftover after biogas production can be used as bio-fertilizer for foliage feedstock production for animals.

In a sustainable livestock farming system one can monitor the feed conversion efficiency (FCE). It is a biological process to calculate the amount of products in terms of milk, meat, and other derivatives produced per unit of feed supply to an animal [60]. The efficiency can be increased by application of new technologies in converting feed energy to milk or meat energy. The individual consumes feed as source of energy; the source of energy is converted into different products such as meat, milk, and fibre in livestock. The energy gained by an individual is used for body maintenance such as digestion, metabolism, growth, and immune function. After fulfilling the body requirement, the energy is channelized for milk or fibre production.

Livestock farming, especially the farming of ruminant's domestic animals can convert cheap and commonly available plant resource as high quality human-edible food. It is only the ruminants can increase quality food supply by exploiting inedible biomass to produce edible protein [61, 62]. It gives an idea about the food use efficiency. Ruminant systems are more efficient as compared to non-ruminant system on the basis of input and output edible to human [60]. Animal products are relatively rich in high value energetic proteins such as vit-B12, vitamin D, and Zinc. The practice of conventional type of livestock production is in a process of shifting to sustainability intensive practice for more production and food security [63]. About 30–40% of cereals grown globally are used as feed for livestock [64]. Ruminant systems are more efficient when compared with non-ruminant system on the basis of inputs and outputs edible to human [65]. Small ruminants like goat and sheep farming systems are the best way and means for the better utilization of marginal lands with prevalence of pastoral system, low level of mechanization, and production of typical products like cheese, high quality meat, therapeutic milk proteins. Even small ruminants are more efficient feed converters as compared to large ruminants when supplied with coarse forage with low nutrient content.

References

1. Anomaly J (2015) What's wrong with factory farming? Public Health Ethics 8(3):246–254
2. Zeder MA (2012) The domestication of animals. J Anthropol Res 68:161–190

3. Olsen KM, Wendel JF (2013) A bountiful harvest: genomic insights into crop domestication phenotypes. Ann Rev Plant Biol 64:47–70
4. Fuller DQ (2010) An emerging paradigm shift in the origins of agriculture. Gen Anthropol 17:1–12
5. Larson G, Karlsson EK, Perri A et al (2012) Rethinking dog domestication by integrating genetics, archaeology, and biogeography. Proc Natl Acad Sci U S A 109:8878–8883
6. Vigne JD, Briois F, Zazzo A et al (2012) First wave of cultivators spread to Cyprus at least 10,600 y ago. Proc Natl Acad Sci U S A 109:8445–8449
7. Serjeantson D (2009) Birds (Cambridge manuals in archaeology). Cambridge University Press, Cambridge
8. Cucchi T, Hulme-Beaman A, Yuan J et al (2011) Early Neolithic pig domestication at Jiahu, Henan Province, China: clues from molar shape analyses using geometric morphometric approaches. J Archaeol Sci 38:11–22
9. Vigne JD (2011) The origins of animal domestication and husbandry: a major change in the history of humanity and the biosphere. C R Biol 334:171–181
10. West B, Zhou BX (1988) Did chickens go north? New evidence for domestication. J Archaeol Sci 15:515–533
11. Fuller DQ (2006) Agricultural origins and frontiers in South Asia: a working synthesis. J World Prehist 20:1–86
12. Nakajima T, Nakajima M, Mizuno T et al (2012) On the pharyngeal tooth remains of crucian and common carp from the Neolithic Tianluoshan site, Zhejiang Province, China, with remarks on the relationship between freshwater fishing and rice cultivation in the Neolithic Age. Int J Osteoarchaeol 22:294–304
13. Fuller DQ, Willcox G, Allaby RG (2011) Cultivation and domestication had multiple origins: arguments against the core area hypothesis for the origins of agriculture in the Near East. World Archaeol 43:628–652
14. Garcea EA (2004) An alternative way towards food production: the perspective from the Libyan Sahara. J World Prehist 18:107–154
15. Peters J, von den Driesch A (1997) The two-humped camel (*Camelus bactrianus*): new light on its distribution, management and medical treatment in the past. J Zool 242:651–679
16. Uerpmann H, Uerpmann M (2002) The appearance of the domestic camel in South-East Arabia. J Oman Stud 12:235–260
17. Bloch G, Francoy TM, Wachtel I et al (2010) Industrial apiculture in the Jordan valley during Biblical times with Anatolian honeybees. Proc Natl Acad Sci U S A 107:11240–11244
18. Clutton-Brock JA (1987) Natural history of domesticated mammals. Cambridge University Press, Cambridge, p 208
19. Diamond J (2002) Evolution, consequences and future of plant and animal domestication. Nature 418:700–707
20. Pringle H (1998) Neolithic agriculture: reading the signs of ancient animal domestication. Science 282:1448
21. Mason IL (1984) Evolution of domesticated animals. Longman, London
22. Zeder MA, Hesse B (2000) The initial domestication of goats (Capra hircus) in the Zagros Mountains 10,000 years ago. Science 287:2254–2257
23. Zeder MA (2006) Central questions in the domestication of plants and animals. Evol Anthropol 15:105–117
24. Taberlet P, Valentini A, Rezaei HR et al (2008) Are cattle, sheep, and goats endangered species. Mol Ecol 17:275–284
25. Taberlet P, Coissac E, Pansu J et al (2011) Conservation genetics of cattle, sheep, and goats. Compt Rendus Biol 334:247–254
26. Felius M, Beerling ML, Buchana DS et al (2014) On the history of cattle genetic resources. Diversity 6:705–750
27. Groeneveld LF, Lenstra JA, Eding H et al (2010) Genetic diversity in farm animals—a review. Anim Genet 41:6–31

28. Marshall FB, Dobney K, Denham T et al (2014) Evaluating the roles of directed breeding and gene flow in animal domestication. Proc Natl Acad Sci U S A 111:6153–6158
29. Scherf BD (2000) World watch list for domestic animal diversity, vol 2000, 3rd edn. FAO, Rome, pp 1–36
30. Schilber J, Elsner J, Schlumbaum A (2014) Incorporation of aurochs into a cattle herd in Neolithic Europe: single event or breeding? Sci Rep 4:5798
31. Larson G, Burger J (2013) A population genetics view of animal domestication. Trends Genet 29:197205
32. Wiener P, Wilkison S (2011) Deciphering the genetic basis of animal domestication. Proc R Soc B Biol Sci 278:3161–3170
33. Kijas JW, Lenstra JA, Hayes B et al (2012) Genome-wide analysis of the world's sheep breeds reveals high levels of historic mixture and strong recent selection. PLoS Biol 10:e1001258
34. Amills M, Capote J, Tosser-Klopp G (2017) Goat domestication and breeding: a jigsaw of historical, biological and molecular data with missing pieces. Anim Genet 48:631–644
35. Librado P, Fages A, Gaunitz C et al (2016) The evolutionary origin and genetic makeup of domestic horses. Genetics 204:423–434
36. Delgado C (2005) Rising demand for meat and milk in developing countries: implications for grasslands-based livestock production. In: McGilloway DA (ed) Grassland: a global resource. Wageningen Academic Publishers, pp 29–39
37. Steinfeld H, Gerber P, Wassenaar T (2006) Livestock's long shadow: environmental issues and options. FAO, Rome
38. FAO (2010) Food and agriculture organization of the United Nations statistical databases. http://faostat.fao.org/
39. GEO4. Global Environment Outlook 4 (2007) Environment for development. United Nations Environment Programme. http://www.unep.org/GEO/geo4
40. Rosegrant MW et al (2009) Looking into the future for agriculture and AKST (Agricultural Knowledge Science and Technology). In: McIntyre BD, Herren HR, Wakhungu J, Watson RT (eds) Agriculture at a crossroads. Island Press, Washington, pp 307–376
41. Herrero M et al (2009) Drivers of change in crop–livestock systems and their potential impacts on agro-ecosystems services and human well-being to 2030. Study commissioned by the CGIAR System wide Livestock Programme (SLP). ILRI, Nairobi
42. Bruinsma J (2003) World agriculture: towards 2015/2030, an FAO perspective. Earthscan, FAO, Rome
43. McIntire J, Bourzat D, Pingali P (1992) Crop-livestock interactions in sub-Saharan Africa. World Bank, Washington
44. NRC (National Research Council) (2009) Emerging technologies to benefit farmers in sub-Saharan Africa and South Asia. National Academies Press, Washington
45. Vallat B, Acar JF, Schudel A (2005) Antibiotic use in animal production and consequences on food safety. In: WAAP Book of the Year 2005. Wageningen Academic Publishers, pp 311–317
46. Smith P, Martino D, Cai Z et al. (2007) Agriculture. In Climate change 2007: Mitigation. In: Metz B, Davidson OR, Bosch PR et al. (eds.) Contribution of working group III to the fourth assessment report of the intergovernmental panel on climate change. Cambridge University Press, Cambridge
47. Seré C, Steinfeld H (1996) World livestock production systems: current status, issues and trends. FAO Animal Production and Health, FAO, Rome, p 127
48. FAO (2017) FAOSTAT. Food and Agriculture Organization of the United Nations, Rome
49. Sere C, Steinfeld H (1995) World livestock production systems. Current status, issues and trends. FAO, Rome
50. Solaiman S (2010) Goat science and production. Blackwell Publishing, Ames
51. Devendra C, Morton JF, Rischkovsky B (2005) Livestock system. In: Owen E, Kitalyi A, Jayasura N, Smith T (eds) Livestock and wealth creation. Nottingham University Press, Nottingham, pp 29–52

52. IPCC (2013) Summary for policymakers. In: Stocker TF, Qin D, Plattner GK, et al (eds) Climate change 2013: the physical science basis. Contribution of working group I to the fifth assessment report of the intergovernmental panel on climate change. Cambridge University Press, Cambridge. p 1535
53. Beauchemin KA, McAllister TA, McGinn SM (2009) Dietary mitigation of enteric methane from cattle. CAB reviews: perspectives in agriculture, veterinary science. Nutr Nat Resour 4:035
54. Ellis JL, Bannink A, France J et al (2010) Evaluation of enteric methane prediction equations for dairy cows used in whole farm models. Glob Chang Biol 16:3246–3256
55. Martin C, Morgavi DP, Moreau D (2010) Methane mitigation in ruminants: from microbe to the farm scale. Animal 4:351–365
56. Grainger C, Beauchemin KA (2011) Can enteric methane emissions from ruminants be lowered without lowering their production? Anim Feed Sci Technol 166-167:308–320
57. Nolan JV, Hegarty RS, Hegarty J et al (2010) Effects of dietary nitrate on rumen fermentation, methane production and water kinetics in sheep. Anim Prod Sci 50:801–806
58. Raman S (2006) Agricultural sustainability: principles, processes, and prospects. Food Products Press, An Imprint of The Haworth Press, Binghamton
59. FAO (2011) World livestock 2011 - Livestock in food security. FAO, Rome
60. Reynolds MP, Hellin J, Govaerts B et al (2012) Global crop improvement networks to bridge technology gaps. J Exp Bot 63(1):1–12
61. Gill M, Smith P, Wilkinson JM (2010) Mitigating climate change: the role of domestic livestock. Animal 4:323–333
62. Wilkinson JM (2011) Re-defining efficiency of feed use by livestock. Animal 5:1014–1022
63. Gerber PJ, Vellinga TV, Steinfeld H (2010) Issues and options in addressing the environmental consequences of livestock sector's growth. Meat Sci 84:244–247
64. Godfray HCJ, Beddington JR, Crute IR et al (2010) Food security: the challenge of feeding 9 billion people. Science 327(5967):812–818
65. Sands M, McDowell RE (1978) The potential of the goat for milk production in the tropics. Cornell University, Department of Animal Science, Ithaca, p 60

Mass Balance Concept in Livestock Farming 2

Mass balance is also known as material balance. It is a practice of accounting for matter entering and leaving into a system which may be helpful in understanding the conservation or steady state of the system under analysis [1]. The exact conservation law used in the analysis of the system depends on the context of the problem but limited to mass conservation, i.e. that matter cannot be destroyed or be created spontaneously. The basic principles of mass balance are applicable to understand about a system stability or steady state, mostly in the field of engineering and environmental analysis. Generally, the principle of material balance can be applicable to design chemical reactor or to understand the stability of an ecosystem under variable conditions. The principles of mass balance are also applicable to livestock production system management in order to bring sustainability in the system.

Mathematically the mass balance for a system is expressed as:

Material (s) input = (material accumulated + material output).

Let us consider the basic unit of livestock production, i.e. animal flock or livestock farmhouse in which a definite number of animal (cattle) are grown at initial stage and maintained further (Fig. 2.1). In this example, there are two units within the boundary layer of the system. The number of animal and provision of outlet for animal wastes and the increase number of animal with the passing of time. The assumptions for the model are steady state, active system within boundary and analysis.

Suppose that the feed, water, and energy input (composition by mass) is "e" kg solid feed (foliage, grain) with "v" of water, with the mass flow of e + v kg/day. The livestock growing within the boundary line should be maintained at steady state or in sustainable growth pattern by keeping the waste accumulation at zero level or minimum quantity, so input and output must be equal for both wastes (manure and wastewater). The output of manure, wastewater, and animal proteins can be monitored by regulating the input of feed (on the basis of consumption per animal).

So, the sustainable pattern of growth of number of animal within the boundary layer of the system mainly depends on the timely input of feedstock, energy (in term of proteins nutrients), and water in order to maintain zero or minimum accumulation

P. K. Rout, B. K. Behera, *Sustainability in Ruminant Livestock*,
https://doi.org/10.1007/978-981-33-4343-6_2

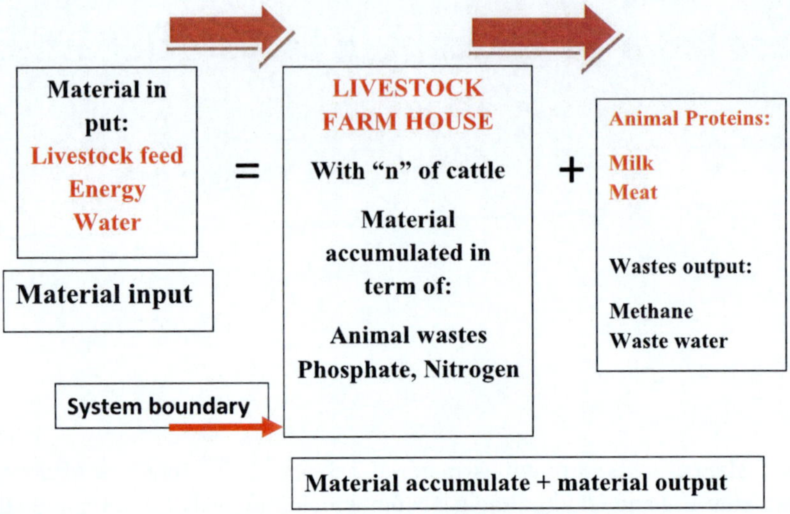

Fig. 2.1 Material balances in a livestock production system

and maximum animal protein output. The wastes "output" from the system should be used as "input source" energy for generation of methane gas and bio-fertilizer for livestock feeds (cereals). In case of misuse of wastes output may cause instability of the farm house in long run. The mass balances of input of materials are discussed in detail.

2.1 Phosphorus Flow Across Livestock Production System

Phosphorus (P) is an essential element for animal and plant growth and development, and is helpful in profitable crop and livestock production. The excess of phosphorus may be harmful for biological productivity, and responsible for accelerating eutrophication in natural aquatic ecosystem [2–4]. So, it is necessary to monitor phosphorus balance in ecosystems, especially in agricultural and livestock farming system. But, proper balancing of rations and management of feed resources has shown positive effect on whole-farm P balance [5–7]. Proper balancing of rations and management of feed resources has shown positive effects on production system (livestock farm) P balances [7, 8]. It has been noticed that the milk production can be increased by regulating the quantity of P in the feed [7, 8]. In addition, the additional P input also causes improvement in the quality of milk production [3].

Phosphorus recycling process in natural ecosystem is presented in Fig. 2.2. Phosphorus becomes available to the plants/crops in soluble form. The soluble form of phosphorus is the key form of available phosphorus for plant growth and development. Due to low concentration of soluble phosphorus, it is necessary to replace as many as 500 times during growing season of plants and crops. The bulk of

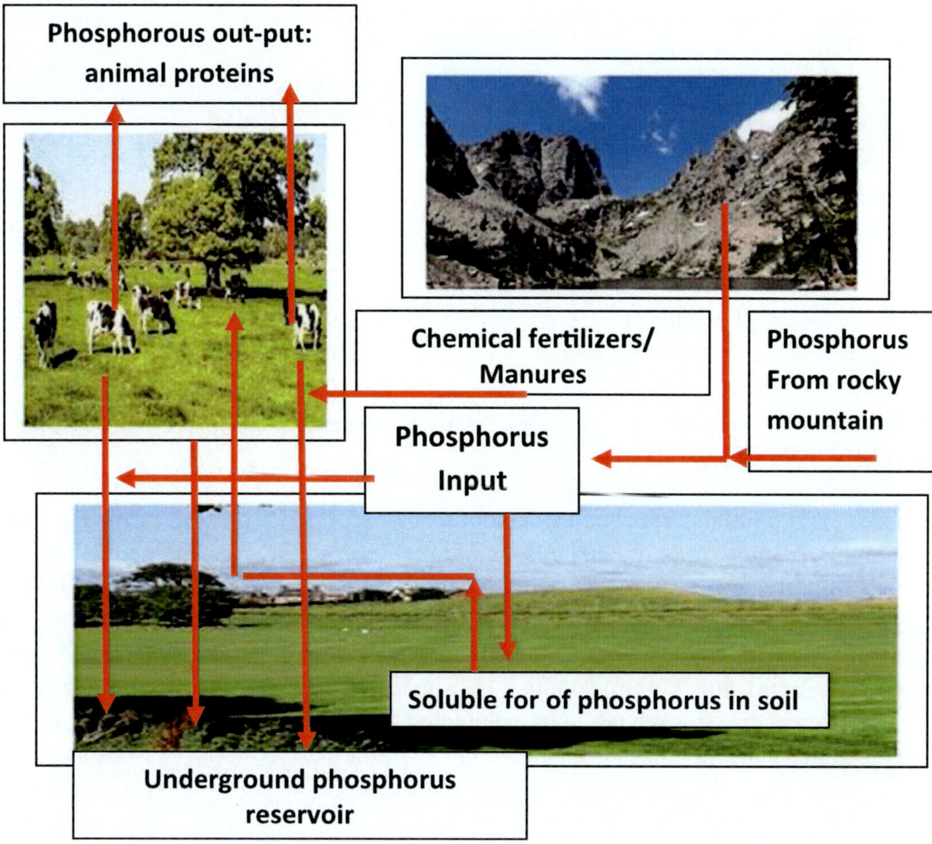

Fig. 2.2 Phosphorus recycling process in natural ecosystems

the soil phosphorus is either in the soil organic matter or in the soil minerals. The phosphorus content in these fractions is highly stable, and present in the soil as unavailable form. The phosphorus added to soil in the form of fertilizer or manure gets bind to the soil strongly and becomes unavailable to crops or trees. So it is difficult to monitor phosphorus in the soil and its availability to plant systems.

In livestock production system, the mass balance for phosphorus is calculated as the difference between input and output of mass across the production system. Generally, by means of mass balance, nutrient management planning in a livestock farming system is monitored to increase the efficacy of animal production. Moreover, the feed management can be helpful in reducing the additional nutrients into the farm and reduce the excretion of nutrient in manure. Generally, manure from ruminants is a primary source of nitrogen and phosphorus for soil surface and groundwater. Manure runoff from cropland and pastures or discharging animal feeding operations and concentrated animal feeding operations (CAFOs) often reaches surface and groundwater system through surface runoff or infiltration. It has been observed that the environmental risk to surface and ground waters is increased if the amount of P imported into the farm (in the form of fertilizer,

Table 2.1 Annual input of phosphate for different livestock farms to maintain zero phosphate mass balance

Livestock Enterprise	Pounds P_2O_5	Acres needed
Growing-finishing beef	17,500	350
Horses	22,000	440
Lactating dairy cows	86,000	1720
Laying hens	1200	24
Cow-calf beef	48,000	960
Sheep	13,500	270
Swine breeding herd with phytase	37,000	740
Swine growing-finishing with phytase	3600	72
Turkeys with phytase	1300	26

With courtesy from: lpelc.org; Livestock Poultry Environmental Learning Community, 2019 by LPELC ADMIN

feeds, and animals) exceeds the amount of P exported from the farm (e.g., crops, animals, manure, milk, meat. eggs, and fibre). The nutrient management planning for farms involved in livestock cultivation depends on the type of domestic animal to be grown for milk, meat, eggs, and other commodities that supplement dietary requirement. Mass balance estimates vary among production system depending upon specific inputs. For example, annual phosphate (P_2O_5) excretion and acreage needed for different livestock have been calculated based on 1000 head/acre, to maintain zero P mass balance (Table 2.1) [9–15].

2.1.1 Monitoring P Mass Balance

Monitoring the flow of phosphorus is entirely different from the strategies for nitrogen across the livestock production system (farmhouse). Phosphorus is strongly held by soil through chemical reaction. In acidic soil the iron and aluminium hydroxide present in the soil hold phosphorus strongly and prevent phosphorus from leaching. In case of calcareous soil, with higher pH (>7), the phosphorus is retained by reacting with calcium. The phosphorus present in the soil can be monitored by means of timely soil test. The phosphorus deficient soil needs addition of phosphorus in time bound manner. The amount of phosphorus to be added in the soil should be well calculated, on the basis of its availability in the soil. The rate of phosphorus accumulation in the soil depends on several factors, such as soil type, rate of phosphorus application, and phosphorus removal due to plantation. Timely soil testing is the only parameter for maintaining plant-available phosphorus in the soil.

The input of phosphorus into livestock Production system (farmhouse) is in the form of feed, manure, bio-solids, and fertilizer, while the output of phosphorus occurs in the form of products such as animal proteins (milk and meat) and forages. The input minus output is referred as "mass balance", which can be expressed on a farm, field, or regional scale. In case of positive mass balance, a production system

Table 2.2 Typical crop nutrient removal for phosphorus

Crop (units)	Per unit of yield P_2O_5	Typical yield per acre	Removal for given yield LB/A P_2O_5
Corn (bU)	0.4	125 (bu)	50
Corn silage (T)[a]	5.0	21 (T)	106
Alfalfa (T)[b,c]	15.0	5 (T)	75
Cool season grass (T)[b,c]	15	4 (T)	60
Wheat/ Rye (bu)[d]	1.0	60 (bu)	60
Oat (bu)[d]	0.9	80 (bu)	70
Barley (bu)[d]	0.6	75 (bu)	45
Soy bean (bu)	1.0	40 (bu)	40
Small grain silage (T)[a]	7	6 (T)	40

[a]65% moisture
[b]For legume grass mixture use the predominant species
[c]10% moisture
[d]Include straw

Table 2.3 Average manure P analyses

	Amount of phosphorus
Dairy	
Lactating cows	4 lb/ton or 13 lb/1000 gal
Dry cow	3 lb/ton
Calves and neifers	2 lb/ton
Poultry	
Boilers	75 lb./ton
Layers	55 lb./ton
Turkeys	80 lb./ton
Swine	
Gestation	35 lb./ 1000 gal
Lactation	20 lb./ 1000 gal
Nursery	40 lb./ 1000 gal
Farrow feeder	35 lb./ 1000 gal
Grew finish	55 lb./ 1000 gal

Actual analyses vary considerable from farm to farm

import more phosphorus than the output values. This is mainly due to either through the direct applications of fertilizer or through the feed phosphorus passing through animal. Unlike nitrogen, there is no gaseous form of phosphorus, so applied phosphorus will build up in soils. The average value for crop removal of phosphorus is given in Table 2.2.

For calculating mass balance for phosphorus flow across the livestock production system (farmhouse), it is also necessary to understand average value for nutrient concentrations in manure (Table 2.3), and average value for nutrient concentration in the livestock animals. The amount of phosphorus at animal unit level is also used in

Table 2.4 Animal unit conversions

Animal	Animal unit	Amount of N (lbs)	Amount of P_2O_5 (lbs)
Layer	250	298	209
Broiler	250	298	209
Swine	5	150	118
Dairy	0.7	16.9	29.3

Source: Kellog et al. [16]

The conversion limits for phosphorus at animal unit level vary from country to country. In Europe, the number of manure-producing animals per hectare of land is limited to 2 dairy cows, 4 beef cattle, 16 fattening pigs, 5 sows with piglets, 100 turkeys or ducks, 133 laying hens, or 285 ground hens

calculating mass flow of phosphorus across the livestock production system (farmhouse). The amounts of phosphorus in animal wastes vary with the type of animal unit. The amount of manure produced from cow is not equivalent to the amount of produced by other animals. Commonly, one cattle phosphorus waste is equivalent to five pigs, and 250 broilers or layers (Table 2.4).

The conversion limits for phosphorus at animal unit level vary from country to country. In Europe, the number of manure-producing animals per hectare of land is limited to 2 dairy cows, 4 beef cattle, 16 fattening pigs, 5 sows with piglets, 100 turkeys or ducks, 133 laying hens, or 285 ground hens. Animal unit coefficient is used as standards for animals based on size and manure production. The animal unit (AU) is a standard unit used in calculating the relative grazing impact of different classes of livestock. One animal unit is defined as 450 kg beef cow (1000 lb) with or without nursing calf, with a daily dry matter forage requirement of 11.8 kg.

Mass balance of nutrients (P_2O_5) across a livestock farmhouse is expressed the difference between input of mass and output of mass across farm boundary (Fig. 2.1). Data analysis on mass balance can provide critical information for nutrient management planning and to manage the movement of nutrient and manure. Environmental risk to surface and ground waters is increased if the amount of P imported into the farm (e.g. from fertilizer, feeds, and animals) exceed the amount of P exported from the farm (e.g. animal proteins, wastes, eggs, animal manure, and fibre).

2.2 Nitrogen Monitoring in Livestock Production System

Nitrogen has significant role in livestock management process and ecosystem management (Fig. 2.3). Nitrogen is one of the important basic requirements for production of animal proteins (tissue, milk, eggs, and wool) the conversion ratio of plant protein into animal protein is illustrated in Fig. 2.4. Grasses as fodder provide more than 70% of the global protein intake by animal.

The conversion of plant protein into animal protein depends on the nature of livestock. About 5–45% of the nitrogen (N) in plant protein is converted to and deposited in animal protein. Mostly, animal system extracts about 55–95% nitrogen

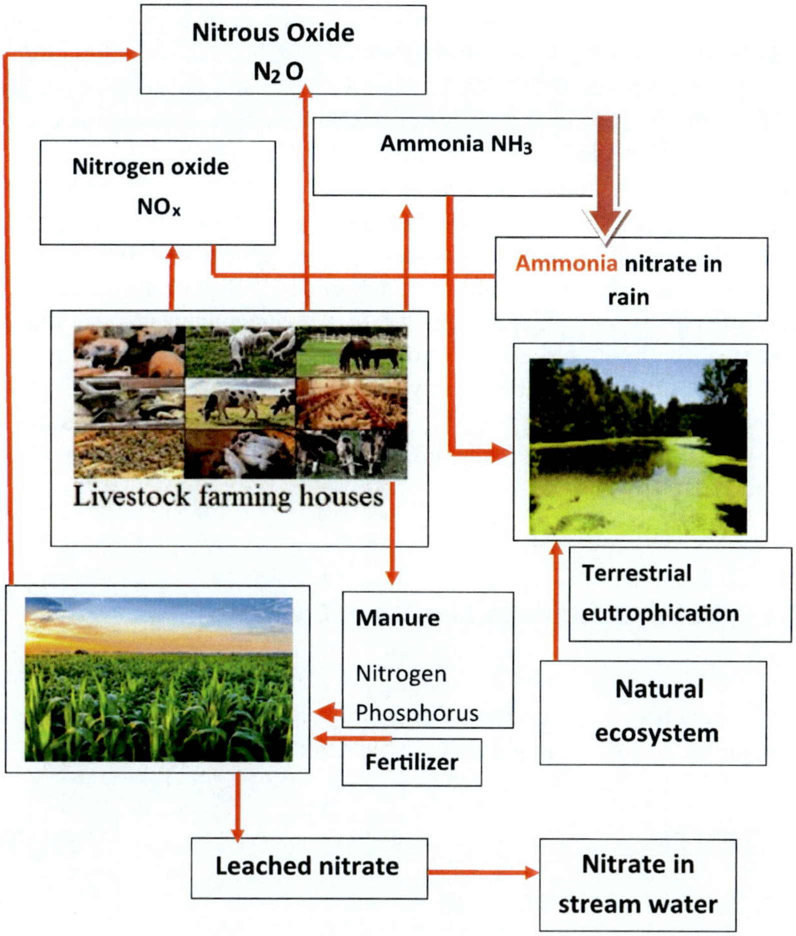

Fig. 2.3 A simplified view of the human impact on the nitrogen cycle and the associated cascading effects

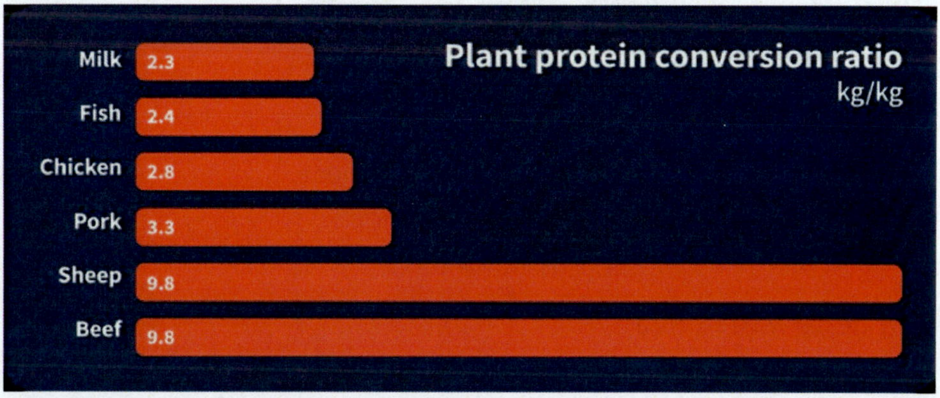

Fig. 2.4 Conversion of plant protein into animal protein

via urine and faeces, and is ultimately used as nutrient for plants. Globally, about 130 TgN per year is produced from livestock which is equivalent to global annual N fertilizer consumption. Cattle (60%), sheep (12%), and pigs (6%) have largest share in animal manure N production. The efficacy of plant N_2 into animal N_2 is noticed to be more in poultry and pork as compared to dairy production. The conversion of plant nitrogen into animal nitrogen is mainly depends on the genetic potential of animals. The conversion process of N manure from animal to plant protein is about 0–60%. The remaining nitrogen is lost to environment via NH_3 volatilization and other process like denitrification, leaching, and runoff in pastures. The N_2 loss from animal manure mainly depends on livestock management process like production potential of the herd, and composition of the animals feed. The efficacy of proper utilization of nitrogen can be improved by crop and livestock production. In addition, by clustering animal production system on the basis of concepts from industrial ecology with manure processing is another possible way to bring amendment in efficacy of N_2 utilization.

2.3 Water Balance in Livestock Farming

Water is a basic requirement in running a livestock production system. Quantity of water input across the boundary of a farmhouse on both annual and daily basis is an important factor for running a sustainable farmhouse (Fig. 2.5).

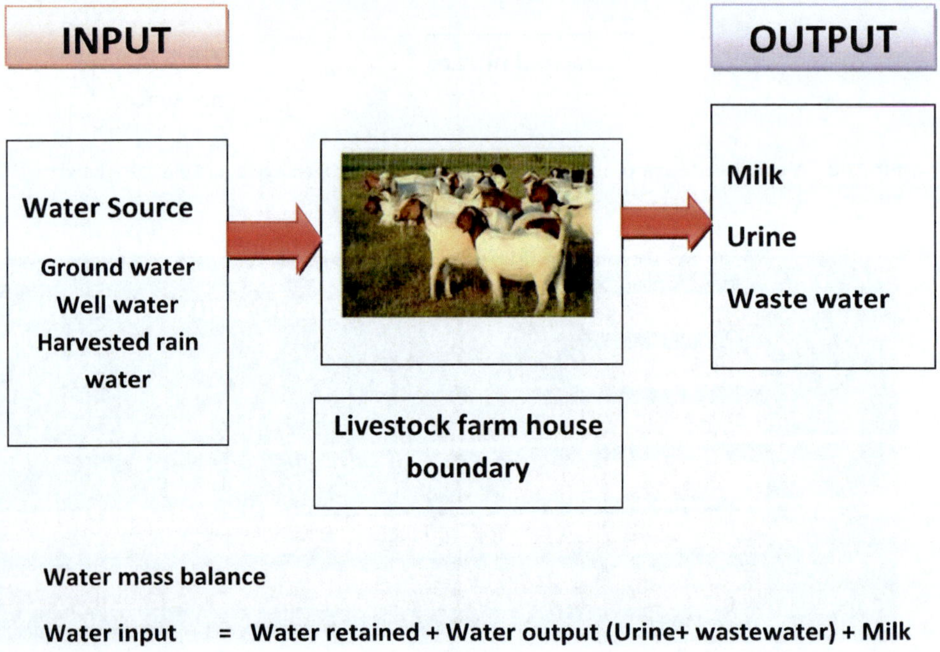

Fig. 2.5 Water mass balance across the livestock farming

The main target of water management is to specify and ensure the guarantee of available water resource throughout the livestock production. So, it is necessary to plan the water input system and associated factors involved in it.

- To understand the farm water requirement, by calculating mass flow balance.
- Reliability of water source availability.
- Arrangement for proper restoration of water on requirement basis.
- To design recycling of wastewater output from the boundary of livestock system.
- To ensure conservation of water during prolong dry conditions.

The water requirement for livestock animal on daily basis varies widely according to the type of animals and their respective activities. It is also affected by factors such as climate, environmental conditions, and the types of feed available to the animals.

2.4 Energy Balance Across the Livestock Production System

Mass flow of energy across the livestock farm house can be expressed in two forms:

1. Precise energy flow through individual animals grown within the bounder layer of the livestock production system.
2. Mass flow of energy across the boundary of livestock farmhouse and its output.

It has been noticed that during early lactation period decrease in animal health and reproduction occur. This is associated with negative energy balance in the animal. So it is necessary to monitor livestock energy balance for maintenance of animal health, reproduction, and feed management.

Energy balance work sheet, in traditional format, for livestock animals (cows) is usually prepared as energy input (EB input) minus output (EB in out) in term s of milk, maintenance, activity, growth, and pregnancy. But the main problem in estimating the energy output (except milk) for other facts is difficult in practice. So, the alternate way of calculating energy output is by body reserve changes (EB body) using body weight (BW) and body condition score (BCS) [17–20]. In order to bring sustainability in livestock system, it is necessary to assess the energy balance for individual cows. Change in body weight is an easy way to understand precise energy flow in cows grown within livestock production system (Fig. 2.6).

Changes in BW and BCS give instant information about changes in body protein, body lipid, and EB body during lactation period. The feed energy content (FEC) is an important factor for overall maintenance of EB out (milk and other animal proteins). The expression of EFC can be presents as:

$$\text{EFC} = (\text{EB body} + \text{E Milk} + \text{E Maintenance}),$$

where EB body = body energy; E Milk = energy in milk; E Maintenance = energy required of maintenance other activities.

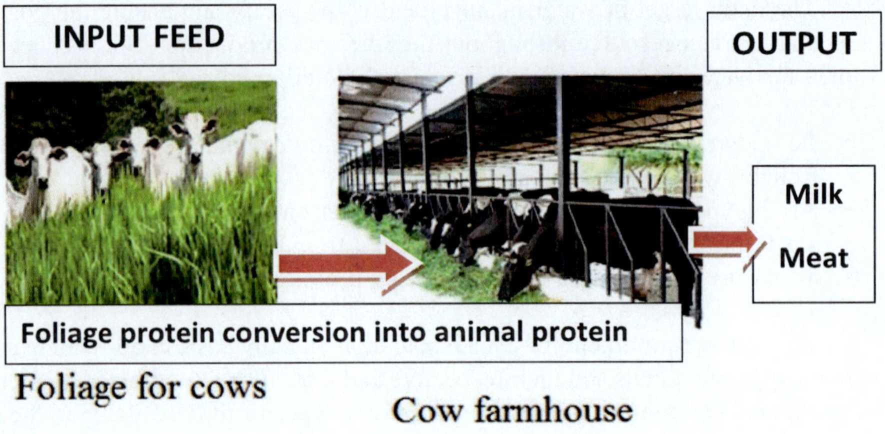

EFC = (EB body + E Milk + E Maintenance),

Where EB body= body energy; E Milk= energy in milk;
E Maintenance= energy required of maintenance other activities;
The feed energy content (FEC)

Fig. 2.6 Precise energy flows across the boundary of livestock farming

2.4.1 Mass Flow of Energy Across Livestock Farming

Without external energy input system, it is difficult to run livestock housing activities like ventilation of air, lighting through electricity, pumping of underground water supply system, air conditioning, etc. It is advisable to exploit natural available renewable energy around the livestock farmhouse to meet day-to-day energy requirement for the utility section of housing systems. Utilization of solar energy, wind energy, and biogas (methane) are few options as energy resource to run the housing system of livestock farm house (Fig. 2.7) [21–25].

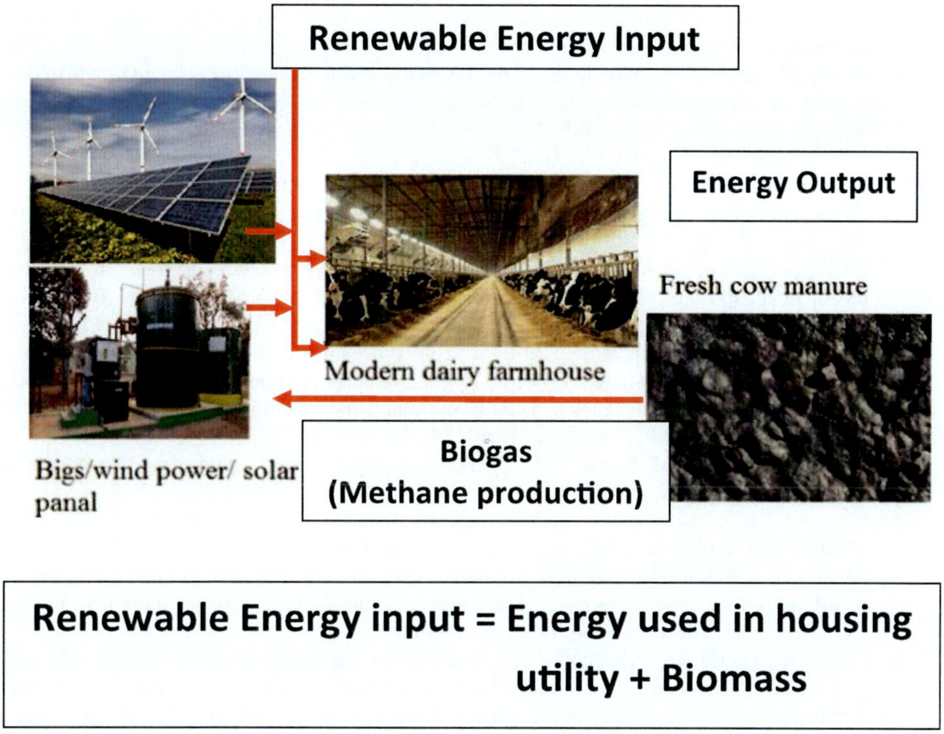

Fig. 2.7 Renewable energy input and utilization in livestock production system

References

1. Himmelblau DM (1967) Basic principles and calculations in chemical engineering, 2nd edn. Prentice Hall
2. Sharpley AN, Chapra SC, Wedepohl R et al (1994) Managing agricultural phosphorus for protection of surface waters: issues and options. J Environ Qual 23:437–451
3. Sharpley AN, Daniel TC, Sims JT et al (1996) Determining environmentally sound soil phosphorus levels. J Soil Water Conserv 51:160–166
4. Shepherd R (2000) Nitrogen and phosphorus management on Wisconsin farms: lessons learned for agricultural water quality programs. J Soil Water Conserv 55:63–68
5. Morse D, Head HH, Wilcox CJ et al (1992) Effects of concentration of dietary phosphorus on amount and route of excretion. J Dairy Sci 75(11):3039–3049
6. van Horn HH, Wilkie AC, Powers WJ et al (1994) Components of dairy manure management systems. J Dairy Sci 77:2008–2030
7. Wang SJ, Fox DG, Cherney DJR et al (2000) Whole-herd optimization with the Cornell net carbohydrate and protein system. III. Application of an optimization model to evaluate alternatives to reduce nitrogen and phosphorus mass balance. J Dairy Sci 83:2160–2169
8. Spears RA, Kohn RA, Young AJ (2003) Whole-farm phosphorus balance on western dairy farms. J Dairy Sci 86(2):688–695
9. Bilotta GS, Brazier RE, Haygarth PM (2007) The impacts of grazing animals on the quality of soils, vegetation and surface waters in intensively managed grasslands. Adv Agron 94:237–280

10. National Research Council (2001) Nutrient requirements of dairy cattle, 7th rev. edn. National Academy of Science, Washington
11. Ruddy BC, Lorenz DL, Mueller DK (2006) County-level estimates of nutrient inputs to the land surface of the conterminous United States, 1982–2001. U.S. Geological survey scientific investigations report 2006-5012
12. U.S. Department of Agriculture (n.d.) Manure management webpage 2.16
13. U.S. EPA (n.d.) State animal feeding operations program status reports
14. US Department of Agriculture (2009) 2007 Census of agriculture: United States summary and state data. 1:AC-07-A-51
15. US Department of Agriculture (2012) National agricultural statistics service, farms and land in farms, final estimates, 2003–2007. See also the table 825 of the U.S. Census Bureau's 2012 statistical abstract, agriculture: farms and farmland
16. Kellogg RL, Charles HL, David M et al (2000) Manure nutrients relative to the capacity of cropland and Pasureland to assimilate nutrients: spatial and temporal trends for the U.S., Natural Resources Conservation Service. U.S. Department of Agriculture, Washington. http://www.nhq.nrcs.usda.gov/land/pubs/manntr.html
17. Azzaro G, Caccamo M, Ferguson JD et al (2011) Objective estimation of body condition score by modeling cow body shape from digital images. J Dairy Sci 94:2126–2137
18. Banos G, Coffey MP (2010) Genetic association between body energy measured throughout lactation and fertility in dairy cattle. Animal 4:189–199
19. Bewley JM, Schutz MM (2009) Potential of using new technology for estimating body condition scores. In: Proceedings 18th annual tri-state dairy nutrition conference, Ft Wayne, IN, Ohio State University, Columbus, pp 23–38
20. Cammell SB, Beever SD, Sutton JD et al (2000) An examination of energy utilization in lactating dairy cows receiving a total mixed ration based on maize silage. Anim Sci 71:585–596
21. Arthur R, Baidoo MF (2011) Harnessing methane generated from livestock manure in Ghana, Nigeria, Mali and Burkina Faso. Biomass Bioenergy 35:4648–4656
22. Currie J (2007) Food, feed and fuels: an outlook on the agriculture, livestock and biofuel markets. Goldman Sachs International, New York
23. Keri B, Thomas D, Kyoung S et al (2008) Livestock waste-to-bioenergy generation opportunities. Elsevier, Amsterdam
24. Kothari R, Tyagi V, Pathak A (2010) Waste-to-energy: a way from renewable energy sources to sustainable development. Renew Sust Energ Rev 14:3164–3170
25. Lin D (1997) The development and prospective of bioenergy technology in China. Biomass Bioenergy 15:181–186

Goat and Sheep Farming

3

3.1 Introduction

Small ruminants are spread all over the world and adapting well to varied agro-ecological regions. Small ruminants are serving the mankind since the dawn of civilization dating to Neolithic Period [1]. It is believed that goat was the first domesticated livestock species to produce edible products for human beings [2]. Small ruminants have proved to be the life saviour for farmers in many catastrophic situations. Goats are particularly very useful and hardy animals and can be productive in the environmental conditions that do not support sheep and cattle rearing [3]. The distribution of goat and sheep varies between economically developed and developing countries. Sheep is almost equally distributed throughout the world. However, the population of goat is more in the low income countries (LRI) than in the high income countries (HRI) [4]. During the early part of the twentieth century, cattle or sheep farming was more popular than goat farming. Subsequently, cattle and sheep farming gained momentum, while goat farming was not popular and goat stock fell due to industrialization of agriculture. However, during the second part of the century, the trend reversed and the number of goats around the world increased substantially. The trend of sheep and goat population and production in countries, categorized as low income (LRI) countries, intermediate income (IRI) countries, and high income (HRI) countries indicated that 79.6% of goats were available in LRI, 19% in IRI, and 1% in HRI countries. The trend of sheep distribution showed that about 40% of sheep were in LRI, 34% in IRI, and 27% in HRI countries. Small ruminants have played a significant role for the livelihood security in smallholder system. Small ruminants have fulfilled agricultural, economic, cultural, and religious roles in all the parts of the world. Small ruminants are adaptable to harsh environment and are better choice of animals in arid and extreme humidity. Small ruminants are promising for subsistence production and contribute significantly to family nutrition. Small animals are easy to manage with minimum financial risk and faster return on investment. The space requirement for handling and feeding of small animal stock is less. More importantly, small

P. K. Rout, B. K. Behera, *Sustainability in Ruminant Livestock*,
https://doi.org/10.1007/978-981-33-4343-6_3

animal rearing is economically viable in all types of production systems and is therefore a viable option for the smallholder to rear it and multiply for commercial use. Small ruminant rearing will enhance the farmer's income and provide more employment to women as it is profitable and less vulnerable to shocks with respect to feed requirement and their management. Small ruminants play major role in poverty reduction by providing a better source of income and nutrition to the poor people in the world. They have the potential for delivering a sustainable increase in income and providing high quality and affordable livestock products.

3.2 Small Ruminant Genetic Resources

Goat and sheep breeds are distributed in all the parts of the world. Sheep is domesticated for wool, meat, hide, and manure and goats are providing meat, milk, fibre, skin, and manure. Small ruminants are reared in large number by smallholders and pastoralists in support of livelihoods and food security. A breed covers groups of animals having similar characteristics that depend on geographical area and origin. Small ruminants are known for their production and adaptive diversity in varied climatic and geographical condition. Goat and sheep genetic resources are described with respect to various uses in different production systems.

3.2.1 Goat Breeds

3.2.1.1 Dairy Goat Breeds

Alpine

The Alpine goats are native to Switzerland and known as better milk producing breed. The French Alpine and the British Alpine are descended from Swiss Alpine goats of the Alps. The colour of the breed varies from pure white to different shades such as fawn, grey or brown to black. They may be either horned or polled. Ears are erect and hair is short. The average weight of adult buck and does are 77 kg and 61 kg, respectively. The animals produce high milk yield and average milk yield is 922.5 kg per lactation.

Beetal

The breed is distributed in Gurdaspur area of Punjab (India) and also in Lahore area of Pakistan. The body colour is black and red colour animals are also observed. Males possess beard. The face is convex, lips black, eyes blue-black with white or brownish corneal surroundings. The ear is long pendulous and drooping. The pinna is showing similarity betel leaf in shape. The horn is horizontal with backward and outward twisting. The average body length, height, and heart girth were 86, 92, and 86 cm in bucks and 70, 77, and 74 cm in doe, respectively. The average adult body weight of male is 57 kg and adult female weight is 45 kg. The average lactation yield is 1.2–2.0 kg.

Damascus
The breed is distributed in the Middle East, especially Syria and Lebanon. The breed is locally known as Baladi or Shami, and Aleppo in Turkey [5]. This is a dairy breed that contains some Nubian blood. The face is straight and convex and ears are short. The coat colour is usually red or brown or white. The hair is long, and males have tassels. Both horned and polled types animals are found in home tract.

Damani
The Damani is a milk goat found in the Bannu and Dera Ismail Khan districts in NWF Province, Pakistan. The goats are medium in size with a mature weight of approximately 35 kg. The coat colour is typically black with tan head and legs. The breed exhibit long hair coat and well-developed udders. The average milk production from the does is 1.8 litres per day.

Jamunapari
The Jamunapari goat is a milk producing breed with average body weight of 28.0 kg at 12 months of age and 1.46 kidding rate [6]. It has a majestic white colour coat with long pendulous ear, Roman nose and has trans-boundary distribution [6]. The breed has been used extensively for upgrading local breeds in neighbouring Southeast Asian countries and is an ancestor to the Anglo-Nubian goat.

Jakhrana
Jakhrana goat is distributed in Jakhrana village of Alwar district of Rajasthan (India). The body colour is predominantly black, spotty white characteristic ears and white speckle in muzzle. The breed has horizontal twisting horn. The body weight at 12 months of age is about 24 kg and milk yield is about 1.0–2.0 litre.

Kamori
The Kamori goat is a popular and beautiful breed of goat. It is a milk goat breed which is found in the Sindh province of Pakistan. The goats are dark brown in colour with small coffee-coloured or dark patches over their entire body. They have long and well-developed body. The head is large, Roman nose. Ears are long, wide, and drooping. The tail is small. Male and female both are horned. Udder and teats are well developed. The average bodyweights of the bucks and does are 60 kg and 50 kg, respectively. The average daily milk yield is about 1.5 litres.

La Mancha
The breed has developed in Oregon, USA. The breed was originated in Texas from short-eared Spanish goats and is a docile breed. The breed has short ear and the coat colour varies in different individuals. The breed produces low amount of milk with high fat content.

Laoshan
The Laoshan goat is a dairy goat breed and distributed in Shandong province of China. The Laoshan goat was developed from the selective breeding of local goats

with Saanen goats. The Saanen goat was first introduced to Loushan by German preachers early in 1904. Laoshan goats are usually white in colour with soft and short hairs. Generally polled animals are found in population. The does have well-developed udder. The animals are medium in size breed and the mature bucks weigh about 76 kg and a mature does weigh about 48 kg. The does produce about 557–870 kg of milk per lactation.

Malabari

The breed is also known as Tellicherry goat. The breed is distributed in Kozhikode or Calicut, Kannur and Malappuram districts of North Kerala. This breed is also found throughout Kerala, the parts of Tamil Nadu and Karnataka. The coat colour is white mostly and other colours like black, brown or combination of all these are also observed. Both sexes have small slightly twisted horns directed outward and upward. Ears are medium sized, directed outward, and downward. The average ear length was 16.30 ± 0.20 cm. Males are bearded. Udder is small and round with medium-sized teats small and thin. The goat produces 0.5–1.25 litre of milk per day. The average adult body weights of male and female are 38.5 kg and 30.50 kg, respectively.

Nubian

The breed is called Anglo-Nubian. The breed was developed in England by crossing British goats with African and Indian bucks to form a composite breed in 1870. The breed is used for milk, meat, and hide. The Nubian goats are used for genetic improvement in more than 33 countries [7]. The main coat colour is black (72%) with occasional white and red patches on a black background. The face is convex with long ears, curved horns, and Roman nose. The adult body weights are 80 kg and 61 kg for male and females, respectively.

Oberhasli

The breed was originated in the mountains and valleys of Switzerland. The breed is also known as Swiss Alpine. This is a medium size breed. The body colour is chamoisee colour with black stripes on its face, the base of the ear, belly and legs below the knee. The adult body weight is 55 kg.

Saanen

The colour varies from white to pale cream and has black spots on the nose and udders. They are usually polled and ears are normally pointed facing forward. Mature male weigh about 75 kg and females 50–65 kg. The breed is used to upgrade local goats. The highest recorded milk yield in the tropics is 800 kg in 205 days lactation or 3.9 kg/day.

Sangamneri

The goat is known as Sangamneri goat and distributed in Ahmednagar, Pune, and Nasik districts of Maharashtra, India. This is a medium size goat with white glossy colour. The ear is pendulous and horizontal about 16 cm long. The tail is curved upward and is about 18.45 cm long. The average body weight at 12-month age was 22.50 ± 0.33 kg. The breed is known for milk production and milk yield varies from 0.9 to 1.5 litre per day.

Surti

Surti goat is distributed in Navsari, Surat region of Gujarat, India. This is a medium size goat. The predominant colour is white and sometimes brown hairs are found in different parts of body. The horns are small, running backwards and slightly curved. Ear is medium size and pendulous. The horn length was about 12 cm. Tail is short and erect upward. The body weight at 12 months of age was 21.99 kg. The breed is known for milk production and produces about 80.22 litre during 90 days lactation period.

Toggenburg

This breed is originated in the Toggenburg valley of north-eastern Switzerland. This is one of the oldest diary breeds of Switzerland. The breed is also known as "Alemu". The breeds have white legs, white stripe from nose to eyes and a characteristic white triangle at the back. The hair is short, and the colour is solid from light fawn to dark chocolate. The ears are erect and carried forward. The mature body weight of males varies from 75 to 91 kg and females around 55 kg. Milk production is about 3 litres/day.

3.2.1.2 Meat Goat breeds

Black Bengal

The breed is mainly distributed in West Bengal, Bihar, Odisha, Jharkhand region of India. The breed is known for meat and skin quality. It is a small size and small-legged goat. Chest is wide with deep body. Body looks triangular from side in females, with heavier posterior region. The body colour is mainly black, however, white, brown, and grey colour goats are also observed in breeding tract. The breed possesses a smooth, shiny, and thin coat. Majority of the animals possess straight head, followed by convex and concave type. Ears are erect. The goats have black nose and muzzle. The breed possesses short tail. The average body weight at 12 months of age is 12.92 kg.

Criollo

Criollo goats are distributed in Mexico, Argentina Bolivia, Peru, and Venezuela [8]. They originated during the Spanish conquest and colonization [9]. The breed is adopted for grazing condition in harsh environmental condition. The most common colour is black. Horns are short. The breed is known for meat production but milk production also valued. The average adult body weight is 39 kg.

Chappar

The breed is distributed in the Sind, Baluchistan area of Pakistan. The breed is locally known as Takru, Kohistani, and Jablu. This is a medium size breed and coat colour is black, white or pied. The body weights of male and females are 26 kg and 20 kg, respectively. The breed is used for meat, milk, and fibre.

Kambing Katjang or Pea

The breed is native to Malaysia and Indonesia. It is called Kambing, which means goat, and Katjang, which means beans. The breed resembles the goats found in Myanmar (Burma), Thailand, the Philippines, and Taiwan. The breed is medium in size. The hair is coarse, males have beards, and the coat colour usually is black with some patches. The body weights of adult males and females are 25 kg and 20 kg, respectively [10].

Kiko

The breed is developed in New Zealand from crossing feral goats exhibiting good meat-type qualities with Anglo-Nubian, Toggenburg, and Saanen bucks and is selected for survivability and growth rate in hill environments [11]. The breed is hardy and adapt to harsh environment. The body colour is white or creamy and the breed is highly prolific with good mothering ability. The mature adult does is about 48.6 kg.

Nachi

The breed exhibit dancing walk due to peculiar pastern movement in a partially revolving motion and with heads held high. The breed is distributed in Jhang, Multan Muzaffargarh, Layyah, Bahawalpur, and Bahawalnagar in Punjab province of Pakistan. Nachis are tall in stature almost as tall as Beetal. The coat colour is black but black and white spotted animals are also observed. The head is medium with Roman nose, small and thin horns, medium ears. The females exhibit well-developed udder. The average adult body weight of does is about 60 kg and breeding males weigh more than 80 kg.

Khari

The breed is distributed across the hills and inner valleys in Nepal and can be found in the mid hills and immediate south of high mountains. This is the major breed of Nepal. This is a hardy breed and found in 6 different colour types, namely, Seta (White), Kali (Black), Khari (Brown), Ghorli (Brown mixed with other colours), Singari (Black with white markings), and Dhobini (White with black markings). The kidding rate is 1.6 and kidding interval of 283 days [12]. The animals are small in size and adult body weight varies from 20 to 40 kg.

Mubende

The Mubende goat is an indigenous breed from the Kabale and Bundibugyo districts of Uganda. The coat colour is the mixture of black and white with shiny straight hair. The bucks have manes, and usually without horn. The body weights of adult males and females are 25–35 kg and 22–28 kg, respectively. The breed survives well during drought and can manage without drinking water for few days. The breed is resistant to heart water (a tick-borne disease), worms, and mange.

Maradi or Red Sokoto

The breed is distributed in the Sokoto region of Nigeria. The strain of the goat is also known as Kano brown and Boronu white. The coat colour is red, and skin is valued for leather production. The goats have horizontally positioned short ears. The body weight of adult animals is 20–30 kg.

Nigerian Dwarf

Nigerian dwarf goat is a small size goat and distributed in the region of West Africa. It is a miniature goat and reared as pet animal. The goats are dwarf and colour is black, or chocolate and gold with random white markings. The goats have a straight nose and erect ears. The coat is soft with short to medium hair. The dwarf goats are prolific. The dwarf goat shows are growing in popularity, and the goat is used for show purpose.

Kalahari Red

The breed was developed from crossbreeding between Red Boer and Nubian. The breed thrives well in harsh condition and suitable for desert condition. The body colour is reddish black, drooping ear and horns move backward. The breed has better growth rate and good mothering abilities [13].

Creole

The breed is also known as Reunion Creole. The breed is widely distributed in the West Indies and in 18 different Caribbean and South American countries as trans-border breeds [14]. The goats have short ears and horns. The colour varies from black, brown or pied. The average body weight of males and females are 25 kg and 20 kg, respectively [15].

Savanna

The breed was developed in South Africa from the crossbreeding of indigenous goats with white Boer goats. The coat colour is predominantly white. The animals are well-built with muscular development and strong bones, legs, and hooves [13]. The breed is raised by pastoralists for both milk and meat [16]. The breed is preferred for religious ceremonies and celebration of the birth of a son.

Pygmy

The Pygmy goat is distributed in West African countries. This is a small breed with broad head and short legs. The hair is straight and medium to long. Males have beards. The coat colour is agouti pattern produced by the intermingling of light and dark hairs of any colour [17]. The breed is used as pet animal and used for meat.

Small East African

This is a small size breed and distributed in Kenya. The coat colour varies from pure white to pure black with different shades. The hair is short and fine. The animals have horns which sweep backwards and curving upwards at the tip. Ears slightly droop and males have beards. The average body weights of adult male and females are 32.5 kg and 26.5 kg, respectively.

Boer

The breed has been widely used for upgrading local goats for meat production in more than 48 countries [14]. Boer goats were derived from indigenous goats of Southern Bantu tribes of South Africa and some genes from goats of India and Europe [18]. It is distributed in South Africa. The coat colour is white and has distinctive red or brown heads and a blaze. The breed has long and pendulous ears. The animals have horns. The breed is known for high fertility and fast-growth. The adult body weights for male and females are 110–125 kg and 90–100 kg, respectively. The kidding rate is 200%. The breed is highly prolific with potential for high growth rate and excellent carcass traits.

Sudanese Desert

The breed is distributed in the semidesert area of northern Sudan. The breed is mainly used for milk and meat production. The coat colour varies from white to black and grey colour is also observed. Male and female are both horned. The average body weight of adult males and females are 40–58 kg and 33 kg, respectively. The kidding rate is around 1.2.

Syrian Mountain

The breed is mainly distributed in Syria, Lebanon, and northern Israel [19]. The breed is also known as Black Bedouin, Djelab, Jabali, Jabel or Jebel, or Membrine, Northern Mountain (Israel) Palestinian, and Syrian black. The hair colour is usually black, and ears are long and pendulous. The average body weights of males and females are 60 kg and 40 kg, respectively. The kidding rate is 1.30 and animals are raised for meat, milk, and hair.

3.2.1.3 Fibre Goat breeds

Angora goat

The breed is distributed in Turkey and known for Mohair production. It is small size and very delicate goat. The coat colour varies from red to tans and brown; and some animals are observed as grey and black. The horns are curved to back. The ears are pendulous and legs are short and small. The breed produces 1–2 kg of fibre known as Mohair.

Atlai Mountain

The breed is distributed in Gorno, Atlai, and Siberia, Russia. The breed was developed by crossing Don goats with local animals. The breed is also known for milk and skin production. The breed is well adapted to grazing in high land environment. The coat colour is black with grey undercoat. The goat produces very good quality black colour fibre. The average body weights of adult male and females are 65–70 kg and 41–44 kg, respectively.

Cashmere

The breed is distributed in mountainous region of Ladakh region, India. The breed is also known as Pashmina, Changthangi (Kashmir), Mongolian Alashan Down, Albas Down, Hexi Down, and Tibetan and Xinjian (China). The breed is white with long hair. The horns are twisted, and ears are either erect or horizontal. The animals sheared once a year and produce about 1.1 kg of fleece.

Kurdi

The Kurdi goat is distributed in Kurdistan, Iraq, and Iran. The breed is a cashmere producing breed. Their colour is usually white, black, or brown. They also knew as Karadi, Kurdish, Murgha, Markhaz, Morghose, or Morghoz (Iran). The face is straight and concave and ears are twisted forward [19] and have twisted horns. The adult body weight of males is 50–60 kg.

White Himalayan

The breed is known for fibre production, but its meat is also valued. They are found in the north-eastern Himachal Pradesh, India. The breed is also locally known as Kangra Valley, Chamba, Gaddi, or Kashmiri. They are used as beasts of burden. The colour is usually white. They have long and twisted horns.

Zhongwei

The breed is also known as Chungwei and Zhongwei goats and they are distributed in the arid desert steppes counties in the Ningxia Hui Autonomous Region and Gansu Province of China [20]. The breed is known for cashmere production and pelts from kids slaughtered at 35 days of age. Males and females have horns that twist upward. The average body weight of adult males and females are 39 kg and 24.5 kg, respectively.

Abaza

It is distributed in north-eastern Anatolia. The breed is known for milk production also and average lactation yield is about 200 kg. The coat is made up of short and soft fibres. The body is covered with pink-white fibres with coloured markings around mouth, eyes, and legs. The bucks have long, flat, and sward-like horns. The does are usually polled [21, 22].

Chyngra

The goats are migratory in nature and spread across trans-Himalayan region of Nepal. The breed is known for finest quality of luxurious cashmere fibre. The goats have spirally coiled twisted horns. The goats are white or black with white or coloured stripes in the face. The average adult body weight is 35–40 kg and 27–30 kg for male and female, respectively.

Jining Grey

The breed is known for production of wavy pattern kid pelt and cashmere fibre production. The breed is distributed across the Shandong Province of China. This is a small size goat and coat colour patterns vary between black, white or black and white. Both bucks and does have horn and carry a forelock. The average body weight of the buck is about 33.2 kg and adult does is 25.4 kg. Bucks will produce about 50–150 grams with cashmere fibre being 18–30% of the total fleece.

Nigora

The breed was developed by crossbreeding Nigerian Dwarfs with Angora goats. Their size is medium to medium small. The American Nigora Goat Breeder Association was established in 2007 in the USA. The registered Nigoras produce three types of fibres. The breed produces both mohair and cashmere like fine fibre. The animals are intelligent, playful, and gentle and reared as pets.

3.2.2 Sheep Breeds

Sheep is reared for production of fine wool, long wool, and mutton. The dual-purpose sheep breeds are popular, which produces both wool and meat. Sheep breeds can be divided into three major categories, i.e. Ewe breeds, Ram Breeds and Dual-purpose breeds. Ewe breeds mainly produce fine or medium wool, long wool and have better milking ability and reproductive efficiency. The Rambouillet, Merino, Targhee, Finnsheep, Corriedale, and border Leicester are classified as ewe breed category. Ram breeds are mainly meat producing breeds and known for growth and carcass characteristics. The Suffolk, Hampshire, Cheviot, South down are classified as Ram breed. Similarly, the dual-purpose breeds are used as Ewe breeds or Ram breeds. The Dorset, Lincoln, and Romner breeds are used as Ram breeds to improve milk yield and fertility; similarly, the Dorset and Columbia breeds are used as ewe breed.

3.2.2.1 Merino

The breed is known for producing fine to very fine wool. The breed was originated and improved in Extremadura in Southwestern Spain. The breed is further refined in New Zealand and Australia and known for fine wool production all over the world. The breed thrives well in arid conditions and found in Australia, South Africa and parts of New Zealand. This breed is found in many countries of the world and the quality of fleece produced varies greatly depending on production environment and management practices. This is a medium-sized animals and the growth rate is low to medium and less muscular breed. The typical characteristic is the spiral horns, which grow close to the head. Polled animals are also observed in population. The average body weights of the mature rams and mature ewes are 80–105 kg and 55–80 kg, respectively. The Australian Merino ranges in micron from superfine, 12–13 microns to coarse, 25–26 microns, the finest grown in Australia. The bulk of Merino wool production is 20–23 microns. Staple length varies from 30–90 mm.

3.2.2.2 Leicester Long Wool

The breed has its origin from Britain. This is a medium to large sized, curly hair sheep breed and is known for carcass quality. Leicester is also known as Leicester Longwool. The breed is distributed in the USA, Great Britain, Australia, Sweden, and New Zealand. Leicester is a medium to large breed with a large, high quality carcass. They are white-faced, broad-backed with heavy fleece. The neck is of medium length. The shoulders are strong and level with the back, which is flat. The legs are straight and wide apart and the hooves are black. The fleece is heavy, curly, soft handling and lustrous, with a spiral-tipped staple 200–250 mm in length, 32–38 micron. The fleece generally weighs from 11 to 15 pounds.

3.2.2.3 Rambouillet

It is the largest fine wool sheep breed. The development of the Rambouillet breed started in 1786, when Louis XVI purchased over 300 Spanish Merinos from his cousin, King Spain. The breed was developed in France from the eighteenth century and now found in U.S. Rambouillet is smooth-bodied, horned or hornless sheep. It has white face and white legs. Fleeces of Rambouillet sheep are relatively heavy. The lambs grow rapidly under good feeding conditions to produce satisfactory market weights at from six to nine months of age. The average adult body weight of rams and ewes are 113–135 kg and 68–90 kg, respectively. Mature ewes produce a fleece weight of 8–18 pounds (3.6–8.1 kg) with a yield of 35–55%. The fleece staple length varies from 2 to 4 in. (5–10 cm) and range in fibre diameter from 18.5 to 24.5 microns or 60–80 for the numerical count.

3.2.2.4 Suffolk

Suffolk is a British domestic sheep breed originated in the late eighteenth century in the area of Bury St. Edmunds in Suffolk, as a result of crossbreeding Norfolk. It is a polled, black-faced breed, and is raised primarily for meat. The breed originated from UK and now also found in the USA.

The Suffolks are polled, and have black open faces along with black legs and white-woolled bodies. Suffolks are considered a large breed of sheep. The breed is known for meat production, also good for wool production as well. Suffolk rams are commonly used as a terminal sire on crossbred ewes due to their ability to produce offspring with excellent growth and carcass traits. Mature weights for Suffolk rams range from 250 to 350 pounds (113–159 kg), ewe weights vary from 180 to 250 pounds (81–113 kg). Fleece weights from mature ewe are 5–8 pounds (2.25–3.6 kg) with a yield of 50–62%. The fleeces are considered medium wool type with a fibre diameter of 25.5–33.0 microns and a spinning count of 48–58. The staple length of Suffolk fleece ranges from 2 to 3.5 in. (5–8.75 cm).

3.2.2.5 Cheviot

Cheviot is a dual-purpose breed and distributed in North Northumberland and the areas near Scottish Borders of the United Kingdom, north west Scotland, Wales, Ireland, and the south west England. The Cheviot is a distinctive white-faced sheep, with wool-free head and legs, pricked ears, black muzzle, and black feet.

It is a long-wool breed, hornless and of reasonable frame. Cheviot wool has a distinctive helical crimp, which gives it that highly desirable resilience. Cheviot wool is often blended into other yarns to give resilience and durability to the finished article. The mature body weight of rams is 160–200 pounds (72–90 kg) with the ewes weighing slightly less at 120–160 pounds (55–72 kg). Mature ewes will average a five to ten pound (2.25–4.5 kg) fleece that has a micron measurement of 27.0–33.0 and a spinning count of 48–56.

3.2.2.6 Southdown

The Southdown sheep was developed in Sussex, England during the late 1700 and early 1800s. The breed is distributed originally in England, and also found in New Zealand, Australia, and North America. This is a small to medium-sized breed and mostly polled. The Southdown is adaptable to wet climates. It is an early maturing breed with good lambing ability and average milk production. The adult body weights for Southdown rams range from 190 to 230 pounds (86–104 kg), ewes are slightly smaller and weigh from 130 to 180 pounds (59–81 kg). Fleece weights from mature ewe are between five and eight pounds (2.25–3.6 kg) with a yield of 40–55%. The fleeces are considered medium wool type with a fibre diameter of 23.5–29.0 microns. The staple length of Southdown fleece ranges from 1.5 to 2.5 in. (4–6 cm).

3.2.2.7 Lincoln

The breed is originated from England and known as dual-purpose breed. The breed produces long wool and growth rate is medium and having moderate muscularity. Lincoln, sometimes called the Lincoln Longwool, is the World's largest sheep breed originated from England. It is developed specifically to produce the heaviest, longest, and most lustrous fleece of any breed in the world. It is now one of the Britain's rarer breeds, categorized as "at risk" by the Rare Breeds Survival Trust since there are fewer than 1500 registered breeding females in the United Kingdom. The present-day Lincoln is said to be the result of crossing the Leicester and the coarse native sheep of Lincolnshire. The breed is distributed in different countries such as England, the USA, Australia, Ireland, New Zealand, Brazil, Canada, and several other South American countries. The breed is mainly white colour and shades of black, charcoal, grey, and silver are also observed. Lincolns are rectangular in form, are deep bodied, and show great width. They are straight and strong in the back and cover thickly as mature sheep. Animals are polled. The adult body weight of Lincoln rams was 113–160 kg and mature ewes was 90–113 kg. The staple length is among the longest of all the breeds, ranging from 8 to 18 in (20–46 cm) with a yield of 65–80%. They produce the heaviest and coarsest fleeces of the long-woolled sheep.

3.2.2.8 Dorset

The breed has its origin from England and is also known as dual-purpose breed. The breed is known for high muscularity and has medium to high growth rate. This is also white colour breed. This is among the most popular white face sheep in the world and has a huge presence all over the globe.

3.2.2.9 Polled Dorset

The Polled Dorset breed of sheep was developed for meat purpose at the North Carolina State University. The breed is distributed at UK, USA, and Australia. Dorset fleeces average five to nine pounds (2.25–4 kg) in the ewes with a yield of between 50% and 70%. The staple length ranges from 2.5 to 4 in. (6–10 cm) with a numeric count of 46′–58′. The fibre diameter will range from 33.0 to 27.0 microns. Dorset ewes weigh from 150 to 200 pounds at maturity, some in show condition may very well exceed this weight, rams weigh from 225 to 275 pounds at maturity.

3.2.2.10 Corriedale

It is a dual-purpose breed, and used for both wool and meat. The breed was developed with Merino-Lincoln cross in Australia and New Zealand and first brought to the USA in 1914. The breed is distributed in Australia, New Zealand, the USA, Southern Brazil, Uruguay, and Patagonia. Corriedales are docile, easy care mothers, with high fertility. The breed adapts to wide range of climatic conditions. The animals have broad body with white face and polled. Corriedales produce a thick stapled, bulky fleece, which is popular with spinners and can be used for a range of hand spun garments. The Corriedale produces bulky, high-yielding wool ranging from 31.5 to 24.5 micron fibre diameter. The fleece from mature ewes will weigh from 10 to 17 pounds (4.5–7.7 kg) with a staple length of 3.5–6 in. (9–15 cm). Mature rams will weigh from 175 to 275 pounds (79–125 kg), ewe weights range from 130 to 180 pounds (59–81 kg).

3.2.2.11 Dorper

The breed is adapted to arid climatic condition and has the ability to adjust to varying seasonal changes. The body has a good combination of hair and wool and they vary from medium to large size.

3.2.2.12 Hampshire

It is classified as Ram breed and originated from England. The breed has black face with white wool. The breed has higher growth rate and exhibit high degree of muscularity. The breed produces medium type wool.

3.2.2.13 Suffolk

Those who are interested in having the best type of meat sheep breed must go for Suffolk sheep. They contain medium wool with black legs and face. This is among the largest growing breed, which offers tasty mutton.

3.2.2.14 American Black Bellied/Barbados.

The breed is classified as Ram breed and animals exhibit high degree of muscularity. The breed has originated from England and having white colour with black face. The breed produces medium type wool.

3.2.2.15 Jacob

With impressive horns, dotted monochrome body and face, this hair sheep breed remains popular for its beauty, skin, and appeal.

3.2.2.16 Malpura

Malpura is one of the important mutton purpose sheep breeds and distributed in the semiarid region of Jaipur, Tonk, Swai-madhopur and adjacent areas of Ajmer, Bhilwara, and Bundi districts in Rajasthan. Malpura is a medium to heavy size breed. The animals are fairly well-built, with long legs. Face is light brown. Ears are short and tubular, with a small cartilaginous appendage on the upper side. Both sexes are polled. Tail is medium to long and thin. The fleece is white, extremely coarse, and hairy. Belly and legs are devoid of wool. The body weights at birth, 3, 6, 9, and 12 months of age were 3.27, 17.38, 26.94, 30.04, and 33.57 kg, respectively. The adult annual GFY was 0.935 kg. The age at first service (AFS) and first lambing (AFL) were 503.9 and 675.14 days, respectively.

3.2.2.17 Patanwadi

Patanwadi is mostly a mutton type breed, but it is capable of producing good amount of milk. The breed is distributed in coastal plain region of Saurashtra and Kutch districts, and sandy loamy areas of Patan, Kadi, Kalol, Sidhapur, and Mehsana district of Gujarat. The body size is medium to large with relatively long legs. Animals have typical Roman nose and ears are medium to large, tubular, with a hairy tuft. The animals are polled and produce white fleece which is of medium carpet quality. The average body weights at birth, 3, 6, and 12 month age were 3.60, 18.65, 27.81, and 36.68 kg, respectively. The average first six monthly and adult annual GFY was 0.842 and 1.021 kg, respectively.

3.2.2.18 Muzaffarnagari

Muzaffarnagari is a mutton type heavy breed. The breed is distributed in Muzaffarnagar, Bulandshahr, Sahranpur, Meerut, Bijnor, Dehradun districts of Uttar Pradesh. The animals are medium to large in size. Face line is slightly convex. Ears and face are occasionally black. Both sexes are polled. Ears are long and drooping. Tail is extremely long and reaches fetlock. The fleece is white, coarse, and open. The average body weights at birth, 3, 6, 9, and 12 months age were 3.55 ± 0.02, 16.02 ± 0.12, 24.46 ± 0.18, 29.18 ± 0.19, and 33.52 ± 0.19 kg, respectively.

3.2.2.19 Magra

Magra sheep is also known as Bikaneri, Chokla or Chakri and produces lustrous carpet wool. The breed is distributed in Bikaner, Nagaur, Jaisalmer, and Churu districts of Rajasthan. The animals are medium to large in size. The face is white with light brown patches around the eyes is the characteristic of this breed. Ears are small to medium and tubular. Tail is medium in length. Fleece is of medium carpet quality, extremely white and lustrous and not very dense. The average body weights at birth, 3, 6, and 12 months of age were 3.29, 18.60, 25.18, and 33.26 kg, respectively. The mean adult annual GFY were 1.837 kg.

3.2.2.20 Marwari

The breed resembles black-headed Persian sheep but is smaller in size and has good fleece. The breed is distributed in Jodhpur, Jalore, Nagaur, Pali, and Barmer districts of Rajasthan and the Jeoria region of Gujarat. The animals are medium size with black face, the colour extending to the lower part of neck. Ears are extremely small and tubular. The average body weight at birth, 3, 6, and 12 months of age were 3.24, 17.41, 25.72, and 32.63 kg, respectively. The mean first six monthly and adult annual GFY was 0.495 and 1.201 kg, respectively.

3.2.2.21 Mandya

The breed is also known as Bannur and Bandur. Mandya is a premium mutton type sheep breed and distributed in Mandya district and bordering Mysore district of Karnataka. The animals are small to medium in size with white colour. The body is compact with a typical reversed U-shape conformation from the rear. Ears are long, leafy, and drooping. Coat is extremely coarse and hairy. Evenly placed short and stumpy legs and wide apart hipbones indicated a square type meaty conformation of the breed. The average body weight at birth, weaning, 6, 9, and 12 month were 2.12, 10.74, 15.37, 17.69, and 21.50 kg, respectively. The first six monthly, adult six monthly and adult annual GFYs were recorded as 292.95, 318.89, and 596.67, respectively.

3.2.2.22 Mecheri

The breed is locally known as Mylambadi and Thuvaramchambali in Tamil Nadu. The breed is distributed in Mecheri, Kolathoor, Nangavalli, Omalur, and Tarmangalam areas of Salem district and Bhavani taluk of Coimbatore district of Tamil Nadu. This is a medium sized animals and the body colour is light brown. Body is covered with very short hairs. The skin quality is very good. The average body weights at birth, 3, 6, 9, and 12 months of age were 2.53, 10.77, 14.91, 18.26, and 21.38 kg, respectively.

3.2.2.23 Sonadi

It is a good mutton type breed. The breed is distributed in Udaipur and Dungarpur districts district of Rajasthan and also found in northern Gujarat. The breed is smaller in size with long legs. Ears are large, flat, and drooping. Tail is long and thin. The fleece is white, extremely coarse and hairy. The average body weights at birth, 3, 6, 9, and 12 months of age were 3.11, 11.64, 16.49, 19.99, and 23.61 kg, respectively. The first six monthly, adult six monthly and adult annual GFY were 233.92, 336.83, and 664.78 g, respectively.

3.2.2.24 Changthangi

Changthangi sheep is a potential dual-purpose breed, locally also known as "Changluk". It is famous for its wool and mutton production. The breed is distributed in the Changthang region of Ladakh. The breed is with good fleece cover which has an extraordinarily long staple. The average body weights at birth, 3, 6, and 12 months of age were 2.50, 10.51, 15.68, and 24.52 kg, respectively. The overall GFY was 1.42 kg.

3.2.2.25 Chokla

The breed is locally known as Chapper and Shekhawati. Chokla is fine carpet wool breed and thrives in migration. The breed is distributed in Churu, Jhunjhunu, Sikar, and bordering areas of Bikaner, Jaipur, and Nagaur districts of Rajasthan. This is a medium-sized breed. The face is reddish brown or dark brown. The skin is pink. The ears are small to medium in length and tubular. Tail is thin and of medium length. The average body weight at birth, 3, 6, and 12 months of age was 3.13, 17.14, 24.26, and 31.03 kg, respectively. The mean first six monthly and adult annual GFY were 0.686 and 2.060 kg, respectively.

3.2.2.26 Awassi

Awassi is primarily known for milk, meat, and wool-producing breed in Southwest Asia. The breed is also locally known as Ivesi, Baladi, Deiri, Syrian, Ausi, DuckTales, Nuami or Gezirieh. It is evolved as a nomadic sheep breed through centuries of natural and selective breeding to become the highest milk producing breed in the Middle East. The breed is distributed in Southwest Asia and Syro-Arabian desert. The Awassi sheep breed is mostly found in most of the Middle East countries including Saudi Arabia, Jordan, Iraq, Syria, Lebanon, Israel, Palestine, and Egypt. It is a fat-tailed type and is multi-coloured breed. Animals can also be found with black, white, grey or spotted faces. The fleece is mostly carpet type with a varying degree of hair. The males are horned and the females are usually polled. The ears are long and drooping. It is an extremely hardy breed, well adapted over centuries of use to nomadic and more sedentary rural management under extreme temperature conditions. The average ewe has single lactation yield of over 300 litres (650 pounds) per 210 day lactation. The grease fleece weight is 1.9 kg (4.2 pounds), mean fibre diameter is 33 micron, and staple length is 16.5 cm.

3.3 Production Traits for Small Ruminant Improvement

The foundation of animal breeding research was started from the work of Mendel during 1865. Principles of population genetics developed by Wright [23] in guinea pigs were extended further for application to the breeding of farm animals. Lush [24] developed "Animal Breeding Plans" and his work was later extended to sheep by Hazel [25]. The researchers used the concepts and methods of Lush to develop breeding objectives and effective genetic improvement programme. They estimated heritability and repeatability and developed selection index to carry out genetic improvement programme. Body growth, milk yield, fibre yield, feed conversion efficiency, meat quality traits have been developed by genetic methodology.

Genetic variability of traits is the first criterion to bring improvement in the population and therefore it is necessary to define the desired traits for quantitative expression. The following traits are recorded to bring desired improvement in the flock.

3.3.1 Production Traits

3.3.1.1 Milk Yield Traits

- Yield per day.
- Yield in a defined period of lactation (e.g. 60 days, 90 days, and 140 days).
- Yield per lactation.
- Fat and protein content.
- Total fat and protein production per lactation.
- Lactational milk yield (LMY).
- Fat protein ratio.

3.3.1.2 Growth Traits

- Birth weight.
- Body weight during particular age (e.g. 3, 6, 9, or 12 months).
- Body size (body height, body length, heart growth) over a defined period (e.g. 3–6 or 3–9 months of age).
- Average daily growth (g/day).
- Pre-weaning growth rate.
- Post-weaning growth rate.
- Number of kids born and survived per adult female per year.
- Carcass weight compared to body weight before slaughter.

3.3.1.3 Fibre/Hair

- Amount per clipping.
- Fineness and length.
- Fleece weight.
- Average fibre diameter.
- Mature body weight of female.
- Skin quality.

3.3.1.4 Wool Traits

- Amount per clipping.
- Fineness and length.
- Fleece weight.
- Average fibre diameter.
- Mature body weight of female.
- Skin quality.

3.3.2 Production Parameters in Goats

The diversity in production performance in goats has been presented in Table 3.1. The heritability of different economically important traits in different goat breeds has been reported. The weighted average of direct heritability (ha^2), maternal genetic

Table 3.1 Diversity in the production performance of goats by region of the world

Region/performance	Goats[a]	Mean ± S. D	Minimum	Maximum
Africa				
Body weight (kg)				
Male	32	53 ± 37	20	130
Female	32	39 ± 22	20	94
Litter size (kid)	17	1.4 ± 0.33	1	2.1
Milk yield (kg/lactation)	3	126 ± 116	50	500
Asia and Pacific				
Body weight (kg)				
Male	106	43 ± 17	16	130
Female	106	32 ± 12	14	100
Litter size (kid)	79	1.4 ± 0.37	1	2.9
Milk yield (kg/lactation)	63	136 ± 109	16	550
Europe				
Body weight (kg)				
Male	123	66 ± 13	35	120
Female	124	49 ± 12	24	120
Litter size (kid)	28	1.6 ± 0.35	1	2.2
Milk yield (kg/lactation)	41	299 ± 225	40	775
Latin America and Caribbean				
Body weight (kg)				
Male	10	40 ± 14	15	70
Female	11	30 ± 13	13	50
Litter size (kid)	11	1.4 ± 0.3	1.1	2.0
Milk yield (kg/lactation)	2	63 ± 4	60	65
Near East				
Body weight (kg)				
Male	48	44 ± 14	17	75
Female	47	33 ± 8	15	50
Litter size (kid)	27	1.6 ± 0.43	1.1	2.5
Milk yield (kg/lactation)	32	150 ± 97	35	460
North America				
Body weight (kg)				
Male	5	41 ± 21	22	67
Female	4	35 ± 26	13	60
Litter size (kid)	4	1.0	1	1

Source: Galal [26]

[a]Number of goat breeds with records

Table 3.2 Weighted average of direct heritability (ha^2), maternal genetic effect (hm^2) and common environmental effect (c^2) for growth traits in goats

Traits	n	$Ha^2 \pm SE$	n	$hm^2 \pm SE$	n	$c^2 \pm SE$
Birth weight	52	0.16 ± 0.014	15	0.12 ± 0.013	9	0.09 ± 0.009
3 months weight	40	0.22 ± 0.011	12	0.08 ± 0.011	12	0.09 ± 0.011
6 months weight	22	0.28 ± 0.015	3	0.13 ± 0.010	3	0.08 ± 0.016
9 months weight	15	0.38 ± 0.018	3	0.09 ± 0.009	3	0.10 ± 0.030
12 months weight	21	0.31 ± 0.018	3	0.05 ± 0.024	3	0.05 ± 0.024
18 months weight	3	0.17 ± 0.319				
Pre-weaning daily gain	16	0.17 ± 0.012				
Daily gain from 3–6 months	7	0.28 ± 0.014	3			0.11 ± 0.138
Daily gain from 6–12 months	2	0.03 ± 0.289	1			0.02 ± 0.020

SE pooled standard errors over studies, *n* number of studies from which averages were calculated

Table 3.3 Weighted average of direct heritability (ha^2) and common environmental effect (c^2) for reproductive and milk production traits in goats

Traits	n	$ha^2 \pm SE$	n	$c^2 \pm SE$
Kidding interval	11	0.09 ± 0.010		
1st kidding interval	4	0.002 ± 0.018		
Litter size at birth	16	0.05 ± 0.004	3	0.06 ± 0.015
Litter size at weaning	3	0.06 ± 0.017	1	0.04 ± 0.020
Litter weight at weaning	8	0.04 ± 0.003	2	0.06 ± 0.018
Gestation length	2	0.10 ± 0.010		
Age at first kidding	9	0.17 ± 0.012		
Average daily milk yield	6	0.31 ± 0.015		
90 days milk yield	4	0.23 ± 0.076		
150 days milk yield	2	0.36 ± 0.129		
Lactation milk yield	12	0.33 ± 0.011		
1st lactation milk yield	11	0.43 ± 0.048		
Protein yield/lactation	11	0.40 ± 0.018		
Fat yield/lactation	11	0.36 ± 0.018		
Fat percentage	6	0.52 ± 0.066		
Protein percentage	6	0.54 ± 0.068		
Comp F & P	6	0.41 ± 0.069		
Lactation length	7	0.15 ± 0.035		
Survival rate to weaning	2	0.09 ± 0.020		

effect (hm^2), and common environmental effect (c^2) for growth traits in goats are presented in Table 3.2. The weighted average of direct heritability (ha^2) and common environmental effect (c^2) for reproductive and milk production traits in goats are presented in Table 3.3. The traits for inclusion in selection programme can be taken up based on heritability.

3.3.3 Production Parameters in Sheep

Selection is used as method of genetic improvement to bring genetic changes in economically important traits. Pure breeders have selection programme to produce superior rams. Genetic progress can be made by selection within a breed for traits that are highly heritable and economically important. Higher heritable traits (40% or higher) include mature body size, face cover, skin folds, clean fleece yield, yearling staple length, gestation length, loin eye area, fat weight, and retail cut weight. Traits having low heritability (below 20%) include weaning type score, weaning condition score, multiple births, number of lambs weaned, fat thickness over loin, carcass weight. The heritability of different traits in sheep is presented in Table 3.4.

Commercial producer should assess their flock before designing the breeding system for the flock. The trait of interest should be decided as wool or growth (wool and growth traits tend to be antagonistic). The factors like flock size, potential benefits from heterosis, pasture and management resources should be considered for effective breeding programme. It is also necessary to analyse how breed differences can be utilized to meet the breeding and marketing objectives of the flock.

Market lambs are produced by either from crossing breeds or from mating crossbred ewes with purebred rams, but the key to the success of crossbreeding is the improvement in production traits made by the purebred breeders in their selection programmes. The additive as well as the heterotic effects is evident in crossbred lambs. It is also required to design three- or four-breed crosses, either rotational or terminal for obtaining desired heterosis.

The use of terminal crossing systems is particularly valuable in those production systems where fine wool production is important. For example, in Australia the Merino is the predominant wool-producing breed. While the Merino produces an outstanding quality of wool clip and Merino wethers are lightly muscled and lower in growth rate. In such a system poorer-performing Merino ewes or older ewes in the flock are allocated to the process of producingF_1 females for mating to a terminal sire.

3.3.4 Genetic Defects

The animals should be free from genetic abnormalities and the animals showing genetic defects are usually culled. The usual inherited abnormalities need to be observed and the individuals should be culled. Some of the inherited abnormalities in sheep and goat are discussed below:

Cryptorchidism is the condition in which ram or buck with only one testis descended into the scrotum and should never be used for breeding. The condition is inherited as a simple recessive trait.

Dwarfism is inherited as recessive and is lethal: therefore, ewes and rams producing dwarf offspring should be culled.

Table 3.4 Heritability of production traits in sheep

Traits	Breed	Heritability	References
(A) Body weight and Body weight gain			
Birth weight	Wool breeds	0.21 ± 0.03 (8)	[27]
		0.13 ± 0.04 (6)	[28]
	Dual- purposc breeds	0.19 ± 0.02 (26)	[27]
		0.19 ± 0.02 (19)	[28]
	Meat breeds	0.15 ± 0.02 (6)	[27]
		0.12 ± 0.05 (7)	[28]
	Malpura	0.21 ± 0.04	[29]
	Awassi	0.30 ± 0.04 (3812)	[30]
	Nali cross	0.28 ± 0.10 (914)	[31]
	Multi-breed cross	0.35 ± 0.04 (11943)	[32]
3 months weight	Wool breed	0.23 ± 0.02 (15)	[27]
		0.33 ± 0.03 (15)	[28]
	Dual- purpose breeds	0.18 ± 0.02 (40)	[27]
		0.20 ± 0.01 (42)	[28]
	Meat breeds	0.18 ± 0.04 (7)	[27]
		0.21 ± 0.05 (13)	[28]
	Suffolk	0.21 (2946)	[33]
	Malpura	0.24 ± 0.04	[29]
	Awassi	0.19 ± 0.04 (3251)	[30]
	Lori	0.353	[34]
	New Zealand sheep	0.14 ± 0.002 (1,747,837)	[35]
	Multi-breed cross	0.81 ± 0.05 (10,043)	[32]
6 months weight	Malpura	0.23 ± 0.04	[29]
	Lori	0.345	[34]
	Makooei	0.22 ± 0.07 (1472)	[36]
	Kermani	0.09 ± 0.06 (1054)	[37]
9 months weight	Makuie sheep	0.37 (2767)	[38]
	Makooei	0.28 ± 0.09 (1151)	[36]
	Kermani	0.13 ± 0.08 (765)	[37]
12 months weight	Multi-breed cross	0.49 ± 0.06 (1208)	[32]
	Makuie sheep	0.20 (1948)	[38]
	Kermani	0.14 ± 0.08 (590)	[37]
Weight gain			
Pre-weaning daily gain	Malpura	0.23 ± 0.04	[29]
	Nellore	0.37 (4349)	[39]
	Awassi	0.19 ± 0.04	[30]
	Multi-breed cross	0.20 ± 0.05 (10,043)	[32]

(continued)

Table 3.4 (continued)

Traits	Breed	Heritability	References
Daily gain from 3–6 months	Malpura	0.13	[29]
	Nellore	0.41 (3871)	[39]
Daily gain from 3–6 months	Malpura	0.22 ± 0.04	[29]
	Nellore	0.34 (3871)	[39]
(B) Wool traits			
Greasy fleece yield (GFY)	Bharat merino	0.16 ± 0.01	[40]
Fleece uniformity (FU)	Finnsheep	0.28 ± 0.02	[41]
Density (DS)	Finnsheep	0.38 ± 0.03	
Staple formation (SF)	Finnsheep	0.32 ± 0.02	
Lustre (L)	Finnsheep	0.23 ± 0.02	
Crimp frequency (CF)	Finnsheep	0.45 ± 0.03	
Fineness grade (FG)	Finnsheep	0.43 ± 0.03	
Staple length (SL)	Finnsheep	0.62 ± 0.02	
Fleece weight at 6 month of age	Muzaffarnagari	0.14 ± 0.03 (2807)	[42]
Fleece weight at 12 month of age	Muzaffarnagari	0.16 ± 0.04 (2038)	
Total fleece weight up to 1 year of age	Muzaffarnagari	0.25 ± 0.05 (2038)	
(C) Milk yield traits			
180-d-adjusted milk yield (180 MY)	Crossbreed dairy sheep	0.32 ± 0.04	[43]
Milk composition traits			
180-d-adjusted fat yield (180 FY)	Crossbreed dairy sheep	0.26 ± 0.04	[43]
180-d-adjusted protein milk yield (180 PY)	Crossbreed dairy sheep	0.30 ± 0.04	[43]
(D) Others			
Reproductive traits			
Age at first lambing (AFL)	Kordi sheep	0.18 ± 0.06 (4568)	[44]
	Multi-breed cross	0.04 ± 0.01 (2154)	[32]
Lambing interval (LI)	Multi-breed cross	0.06 ± 0.03 (4600)	[32]
Behavioural traits			
Maternal behaviour score (MBS)	Multi-breed Australian sheep	0.20 ± 0.02	[45]
Lamb vigour		0.40 ± 0.04	[46]
Lamb activity	Mouton Vendeen	0.15 ± 0.05	[47]
	Rouge de l'ouest	0.11 ± 0.0	[47]
Sucking ability		0.32 ± 0.03	[46]
Sucking assistance	Mouton Vendeen	0.24 ± 0.05	[47]
	Rouge de l'ouest	0.22 ± 0.05	[47]
Residual feed intake (RFI)	Romane	0.45 ± 0.08	[47]
Feed conversion ratio (FCR)	Romane	0.30 ± 0.08	[47]

Parrot mouth: If the lower jaw is either too short (overshot or parrot mouth) or too long (undershot), the teeth cannot close against the dental pad and grazing is difficult. The mode of inheritance of these jaw abnormalities is unknown, but the conditions are under genetic control.

Rectal prolapse is common in black-faced sheep. It is a serious defect, and both inheritance and the environment are influential in its occurrence. Lambs on heavy feeding or lush pastures are more likely to show rectal prolapse. If surgery is used to correct the condition, the animal should not be used as a breeding animal. Skin folds, open-faced, closed-faced, and wool blindness are inherited, and selection against these traits has been effective.

Spider syndrome is a severe skeletal deformity occurring in Suffolk sheep. The most striking feature is an outward bending of the forelimbs from the knees. Angular deformities of the hind limbs are generally present. The genetic abnormality is due to recessive gene and can be effectively selected against. Many wool defects are known, including black fibres. Black-tipped fibres, hairiness, fuzziness, and "high belly wool" are some wool defects. Selection against all of these fleece defects should be practiced as a means of reducing the frequency of their occurrence.

3.4 Care and Management of Goats

3.4.1 Age Determination and Body Condition Score

The approximate age of the goat can be determined by observing dentition pattern (teeth development). The age of goat can be estimated based on their eight lower incisor teeth. The first two baby teeth are lost after approximately 1 year of age. Two baby teeth are replaced with permanent teeth in every year and all baby teeth are replaced approximately after 4 years of age (Table 3.5). Similarly the body condition score in goats has been described in Table 3.6.

3.4.2 Management of Breeding Bucks

- Body condition score should be at least 3 out of 5.
- Bucks should be provided supplementary feeding of 300 gm/goat /day for medium breed or 400–500 gm./goat/day for large size breed.

Table 3.5 Age of goats by dentition pattern

Type of teeth	Age (Months)
8 baby teeth (milk teeth	0–14
2 large teeth in the middle (one pair permanent incisor)	14–19
4 large teeth and 4 milk teeth (two pairs of permanent incisors)	19–24
6 large teeth and 2 milk teeth (three pairs of permanent incisors)	24–30
8 large teeth (four pairs of permanent teeth)	30++

Table 3.6 Body condition score for goats

Body condition score		
Condition	Score	Comments
Severely emaciated extremely poor	1 (thin)	Close to death / starvation evident; outline of ribs visible and spinal processes distinct and prominent with severe depressions; physically weak; shoulder, loin and hindquarters atrophied in appearance; skin adhered to bone
Extremely thin poor	2 (thin)	Not weak, not as emaciated as (1); skin in direct contact with bone; prominent "V" shaped cavity under tail; outline of spine and ribs are still visible; bony surface of the sternum protruding
Very thin frame visible	3 (thin)	Wasting in appearance; ribs visible; individual spinal processes evident and depressions obvious (rib, hips) and sunken between pins and hooks; sternum is prominent
Slightly thin	4 (moderate)	Spinous processes(dorsal/transverse) are prominent and sharp; thin flesh covering between hooks and pins; some ribs visible; definite depression between hooks
Frame covered balanced	5 (moderate)	Spinous processes smooth; transverse processes have smooth concave curve; hooks and pins smooth; muscle becoming obvious; sternum can be palpated
Slightly fleshy smooth cover	6 (moderate)	Spinous processes rounded; spinous to transverse processes smooth sloped; hooks and pins covered; slight depression between hooks and pins
Frame not visible fleshy	7 (fat)	No spinous processes noticeable, ribs not visible, hooks and pins rounded with some cover; flat between hooks; palpation of sternum difficult
Obese	8 (fat)	Edge of transverse processes barely noticeable, tail head cavity filling with fat
Extremely obese severe over – condition	9 (fat)	Spinous processes buried in fat; between hooks and pins rounded; between hooks rounded; tail head cavity exhibits fat-filled folds

- All the desired vaccination should be carried out at least 6 weeks before breeding seasons.
- Hooves should be trimmed before breeding seasons for easy mounting and mating ability.
- Foot dipping is recorded during breeding season.
- Deworming for internal parasites should be carried out 3 weeks before to mating.
- Semen characteristics should be analysed before mating.
- Screening for brucellosis and Mycoplasma need to be carried before 30 days of breeding season [48–50].

3.4.3 Age at First Mating

It has been observed that early mating may cause abortion, produce low birth weight kid and may not produce enough milk for kids. The age at first mating of does is dependent on breed and nutritional status of production system. A doe should not be mated before 1 year of age. Ideally a doe should have one pair of permanent incisors which is the age of 14–17 months before first mating.

3.4.4 Heat Detection

Timely detection of oestrus is highly essential for successful mating or AI. It is very difficult to know the exact time point when oestrus started, and therefore very difficult to predict ovulation time. Oestrus monitoring in the flock with an aproned or vasectomized teaser buck twice a day (morning and evening) is essential for identifying oestrous cycling does. It helps to detect cyclic and non-cyclic does. Oestrus symptoms are more distinct in goats unlike cattle and buffalo. This is the optimum time for mating. The ovum remains viable for 10–12 h. Similarly sperm also remains viable for 12–24 h. It is desirable to mate both male and female approximately 20 h of onset of oestrus. An oestrous doe can be detected by observing following symptoms in the flock:-

- Restless, reduced appetite, and milk production.
- Acceptance of mounting of teaser buck.
- Bleats frequently and wagging of tail constantly and rapidly.
- Mounting over other oestrus and non-oestrous females and allowing mounting of other females.
- Clear, transparent sticky mucus discharge from vagina.
- Swollen and reddish vulva.
- Gathering of oestrous females around the teaser buck.

3.4.5 Gestation Period

It is defined as the period of time from conception to parturition in livestock. The gestation period in goat varies between 149 and 150.8 days and average period is 149 days. There are several factors which affect the gestation period. Male kids are carried for longer period than female kids. Similarly, single kids are carried for longer period than twin born kids. The foetal growth about 70% occurs during three last trimesters of pregnancy (about after 100 days) [51, 52]. It has been observed that corpus luteum is essential for the maintenance of pregnancy throughout gestation. It does not happen in the case of ewe where corpus luteum is essential for the first 3 months of pregnancy.

3.4.6 Breeding and Kidding Management

The normal breeding season occurs during September, October, and May–June by which kidding occurs during February and March, and October and November. The records should be kept showing breeding dates and the bucks used. From breeding records, it is possible to calculate when does will deliver kids, as the gestation period is about 150 days.

The bedding with soft dry grass should be prepared in kidding shed or pasture for kidding. The kidding should be carried out in a clean environment and kidding pen should be bedded with straw, soft grass or other absorbent material. The body condition score should be 3.0–3.5 during pregnancy condition. The period 6–8 weeks prior to kidding is very important. The quality feed should be supplied during the period. Clean water and mineral mixture should be available during the period. Screening for brucellosis and faecal egg count should be carried out. A dose of selenium or supplementation should be provided. The sign of impending kidding is udder enlargement, swelling of the vulva, restlessness, loss of appetite, frequent urination, and mucous discharge from the vulva. There are three stages of parturition. The first stage is the uterine contraction and cervical dilation and usually lasts for 2–6 h. The second stage is the abdominal contraction along with uterine contraction and will last from 30 min to 2 h. The doe comes to second stage the water bag will break and if the doe is straining for more than 1 h after the water bag has broken, then assistance is required for kidding. The third stage is the expulsion of placenta and it takes about 5–8 h.

A doe almost ready to deliver young should be placed in a clean pen with soft bedding with clean dry grass or with clean straw. If the kid presents the front feet and head, delivery should be easy without assistance. If only the front feet, but not the head, are presented, the kid should be pushed back enough to bring the head forward in line with the front feet. As soon as the kid arrives, its mouth and nostrils should be wiped to clean mucus. The navel should be kept out of the water. Kids should be encouraged to nurse as soon as they are dry, as nursing helps the new-born kid to keep warm. It is advisable to use disposable rubber or plastic gloves when assisting with delivery and handling everything aseptically.

Naval Chord cutting: This is carried out immediately after kid's birth. The naval cord should be cut 30 cm away from abdomen using sterilized scissors/ blades and should be cauterized by dipping the cut end in tincture iodine solution or povidone iodine. The application of antiseptic should continue till drying and falling down of remnant cord.

Colostrum Feeding: Colostrum is the first milk after kidding which is high in nutrients and antibodies. Colostrum should be fed at the earliest possible time, so that antibodies are absorbed well through kid's intestine to the circulation for protecting the kids from infections at early stages of life.

3.4.7 Weaning Management

Kids need to be provided different types of feed such as the creeper ration, green fodder, and dry fodder gradually. Kids obtain adequate roughage before weaning to promote normal rumen development and proper rumen flora. The castrated and breeding kids should be kept in separate pens. Coccidiosis and pneumonia are the major problems in the post-weaning age group, particularly under conditions of confinement housing. Coccidia commonly resided in intestine and multiplied rapidly in the intestine. The intestinal damage can result in decreased performance, reduced resistance to other diseases, and even death.

Castration: Male kids not acceptable for breeding should be castrated at 4–6 weeks of age. There are 3 methods of castration, viz., Burdizzo's or emasculate method, rubber band or elastrator method and incision or knife method. This can be done by constricting the blood circulation to the testicles by use of a rubber band elastrator or by surgically removing the testicles. Surgical removal of the testicles creates a wound that attracts flies in the warm season; therefore, a fly repellent should be applied around the wound. It is recommended that a veterinarian castrate the older goat. The necessary precaution must be taken to prevent infection, gangrene, and other complications.

Identification marks: Identification is done for maintaining proper records on goat farm, feeding of animals, better management, and insurance of animals. Identification marks are either temporary or permanent. Among permanent methods, tattooing and tagging are used in goats. Tattooing is done with a tattooing machine after fixing the desired number on the inner side of ear (preferably left ear). All kids can be identified by a tattoo in the ear. Ear tags are also useful. The tattoo serves as permanent identification in case an ear tag is lost. All registered goats must be tattooed because the goat registry associations do not accept ear-tag identification. RFID technology is also used for better management of farm operation like data recording and health management.

Bedding: Bedding is mostly provided during kidding and also kids up to weaning. Bedding also prevents kids from licking of soil. The locally available materials like dried waste grasses and straw should be used as bedding. The bedding materials should be replaced daily / rotated to keep it dry. The lime should be used while replacing the bedding materials to keep the floor dry.

Disinfection of Goat housing: To eradicate the pathogenic organisms and for general hygiene of the farm, sanitation programme is must in the farm. Activity related to disinfection / sanitation should be preferably done when goats are out for grazing. For disinfecting the goat shed floor (concrete / brick), splashing with phenyl solution or any recommended antiseptic solution at monthly interval is much effective. In case of kachha floor, removing the upper layer (1–2 cm) of soil and replace it with fresh soil along with lime dusting. Walls should be whitewashed at least in a year. Feeders and waterers and milking equipment should be scrubbed fortnightly with 4% washing soda (sodium bicarbonate) in hot water.

Hoof Trimming: Hoof trimming required in goats to make free from foot problems. Goats' feet should be kept properly trimmed to prevent deformities and foot rot. Hoof trimming should be carried out in every 6–8 weeks. A good pruning shear is ideal for leveling and shaping the hoof, but final trimming can be done with a hoof knife. The heel regions can also be trimmed when needed. The sole may be slightly trimmed till slightly pink colour appears beyond this point bleeding will occur. It is very easy to cut in rainy days where the hooves are soft enough after absorption of moisture.

Brushing/Grooming of Goats: Goats are groomed on standing position on clean floor starting from neck towards the hindquarter in the direction of hair using brush. Sticky hair and dry faeces should be removed before brushing. It helps to become friendlier with animals and stimulates circulation of blood.

Clipping: Clipping is mainly done using hand or electric clippers to maintain cleanliness of hair coat. The long hairs in the hindquarters should be clipped before mating and kidding. Hence it should be hand clipped well in time before commencement of breeding season and during rainy season when soiling of hair is common.

Dehorning: Horned animals should be disbudded during the first week following birth by use of an electric dehorning iron. Goat breeders have been plagued by the genetic linkage between hermaphrodites and polled expression.

Milking: Goats are milked by hand or by milking machines. Clean, sanitary conditions are very important prior to milking. After milking, the teats should be dipped in a weak iodine solution.

3.5 Management Practices of Sheep

The sheep should be handled carefully as they should never be restrained by its wool because the skin is pulled away from the flesh causing a bruise. The best way to hold the sheep is to grasp the skin under the chin with one hand and to grasp the top of the head with the other hand. If a sheep is to be moved forward, then pressure may be applied on the dock which makes the sheep move forward.

3.5.1 Care During Lambing

Ewes should be placed in comfortable housing with soft bedding before lambing. Wool should be clipped from the dock, udder, and vulva regions and the ewes may be completely shorn during comfortable weather. All dung tags (small pieces of dung that stick to the wool) should be clipped from the rear and flank of pregnant ewes. The ewe and new-born lamb should be placed in a lambing cage. In a normal presentation, the head and front feet of the lamb emerge first. If the rear legs emerge first (breech presentation), then the assistance is required. If the front feet are presented but not the head, the lamb should be pushed back enough to bring the head forward for presentation. The mucus should be removed with dry cloth with proper precaution. If a lamb becomes chilled it can be immersed in warm water from the neck down to restore body temperature, after which it should be wiped dry. A lamb that has nursed by ewe will survive without difficulty. The temporary identification mark can be assigned to lamb.

3.5.2 Body Condition Score and Age Determination in Sheep

The body condition score for ewes should be determined and it is an index of individual fitness status and also provides idea about flock management. The body condition score should be determined as per the criteria described in Table 3.7 and desired body condition scores at various production phases is presented in Table 3.8. Simultaneously the farmer should have idea about the age of the animal by observing dentition pattern. Lambs have four pair of narrow incisors called milk teeth or baby teeth. At about 1 year of age, the middle pair of milk teeth is replaced by a pair of large permanent teeth. At about 2 years, second pair of milk teeth is replaced and have four permanent teeth. At about 3 years of age, the sheep possess six permanent incisors. At about 4 years of age, the sheep possess eight permanent incisors.

Table 3.7 Body condition scores for Ewes

Score	Spinous process[a]	Transverse process[a]	General description
1	Very prominent	Clearly protrude	Very thin, no fat cover
2	Prominent, but smooth	Rounded	Thin, but skeletal features do not protrude
3	Rounded, but smooth	Rounded, but smooth	Evidence of fat along rib, shoulder, back; hips protrude
4	Barely evident	Cannot be palpated	Moderate fat deposition throughout
5	Not detectable	Not detectable	Excessively fat

[a]Should be palpated down the spine, along the loin and over the rib

Table 3.8 Desired Body condition scores at various production phases

Production phase	Desired body condition score
Dry ewe (116–176 days)	1.5–2.0
Breeding (35–52 days)	2.5–3.0
Early gestation (first 15–17 weeks)	2.0–2.5
Late gestation[a] (last 4–6 weeks)	2.5–3.0
Early lactation (first 6–8 weeks)	3.0–3.5
Late lactation, weaning (last 4–6 weeks)	2.0–2.5

Source: Adapted from Sheep production handbook [53]
[a]Add 0.5 for ewes pregnant with or nursing twins

3.5.3 Castrating and Docking

The elastrator method can be used by applying a tight rubber band around the scrotum above the testicles (for castration) and around the tail about and 2–3 cm from the buttocks (for docking). However, some complication may arise due to infection and death may occur due to tetanus also. A fly repellent should be applied around any wound to lessen the possibility of fly strike (fly eggs are deposited during warm weather).

Ewes also develop a vaginal or uterine *prolapse* (protrusion of the reproductive tract to the outside through the vulva). This condition is extremely serious and leads to death.

3.5.4 Shearing

Sheep are usually shorn during the spring before the summer. Sheep should be kept off feed and water for 6–12 h prior to shearing to reduce gut fill and to help reduce skin cuts. Preferably the dry sheep should be shorn. A fleece that has been properly clipped will remain in one large piece. It is required to remove dung tags and coarse material from the clipped wool may be stored in a separate container. The core sample is taken by inserting a hollow tube that is sharp at the end into the sack of wool. The sample obtained is examined to evaluate the wool in the sack. The wool should be stored in a dry place and on a wooden or concrete floor to avoid damage from moisture. Sheep should be kept in a dry and clean pens and protection should be provided from cold, wet, windy, or hot conditions Sheep that are nicked during shearing should be treated to avoid infection.

3.5.5 Breeding Soundness

Breeding soundness of the male (buck/ram) for optimal performance is an important aspect of reproductive success. There is numerous aspect of thorough breeding soundness examination i.e., physical examination, and genitalia examination, semen evaluation. Rams/bucks should be tested for brucellosis and other diseases. Mating behaviour can be tested and observed in the field and pen. Low performance males should be identified and removed. Reasons for variation in male performance have been identified as genetic, social, and management related problems.

3.6 Feeding of Small Ruminants

Traditionally sheep and goats are grazing animals and managed in low input system in field condition. It is necessary to adopt controlled feeding for greater nutritional efficiency and to enhance individual animal productivity. The advantage of nutritional supplementation has been demonstrated in developing and developed

Table 3.9 Nutrient requirements of sheep and goats at various life stages

	Sheep		Goat	
Attributes	CP%	TDN%	CP%	TDN%
Maintenance (mature female)	9.6	57.6	10	55
Late gestation	11.2	66.7	11	60
			11[a]	60[a]
Lactation	14.8[b]	64.5[b]	14[c]	65[c]
Early weaned	14.5	75.8	14	68
Finishing (4–7 months)	11.7	77.1	–	–
Yearlings	9.1	57.6	12	65

[a]Average milk
[b]Nursing twins
[c]High milk Source: NRC, 1985; 2007 [54, 55]

countries to improve the small ruminant productivity. The energy and protein requirement of sheep was attempted through feeding trials and extrapolation of cattle data during the period 1908–1933. Fraps and Cory published detailed articles on diet selection for sheep and analysed the nutritional values of forages and feeds. The maintenance energy requirement of sheep was published by NRC in 1957. Subsequently many experiments were carried out to determine more accurately the protein, energy, mineral, and vitamin requirements of sheep in different stages of production. The energy and protein requirements of goats have been described by NRC 2007.

Small ruminants require energy, protein, vitamins, minerals, fibre, and water. The vitamins and minerals are also required for optimum flock performance. Fibre (bulk) is necessary for the rumen fill and to maintain a healthy rumen environment and prevent digestive upsets. Water is the essential requirement for individual as well as flock health.

Feed is the major input associated with livestock farming and accounting for 60% of total production costs for raising small ruminants. Nutrition plays an important role on flock reproduction, milk yield, and weaning requirement of kids and lambs. Similarly, the nutritional inputs play a significant role during gestation and lactation period of ewe and doe.

Small ruminant nutritional requirements can be met by feeding various feedstuffs. The nutritional requirement of sheep and goat is presented in Table 3.9. The primary source of small ruminant nutrient requirement is usually met from pasture, forbs, and browse. Pasture tends to be high in protein and energy when it is in vegetative stage. As pasture plants mature, palatability and digestibility decline, therefore pasture rotation is important for maintaining vegetative state. Goats are natural browser and have the unique ability to select plants when they are at their most nutritious state. Sheep are excellent weed eaters. Dry fodder and concentrate should be provided to meet the energy and protein requirement of animals. Creep feeding should be provided to lambs and kids to sustain weaning growth and to attain finishing weight.

Generally, kids and lambs must be allowed to obtain the first milk (colostrum). Solid feeds are less expensive than milk replacers, and when the kids can do well on solid feeds, milk replacers should not be fed. Kids and lambs should be provided concentrates until their rumens are sufficiently developed to digest roughage to fulfil the body requirement. Goats and sheep should be fed as per age and physiological condition. Dry does and ewes perform satisfactorily on ample browse, good pasture, and good-quality grass and legume hay. Lactating does and ewes should be given good-quality dry fodder and high levels of concentrates. Pregnant does and ewes should be fed to gain weight in order to ensure adequate nutrition of the kids. Young, growing and lactating goats and ewes need more protein than other animals. A ration containing 12–15% protein is desirable for dry animals, but 15–20% protein may be better for young animals and for lactating animals that are producing much milk.

In some areas, deficiencies in minerals such as phosphorus, selenium, and iodine may exist. The use of iodized or trace-mineralized salt along with dicalcium phosphate or steamed bone meal usually provides enough minerals if legume hay or good pastures are available. Calcium is usually present in sufficient quantity in legume hay and phosphorus is adequately provided in grain feeds.

Rams should never be run with doe goats, and buck goats should never be run with ewes. These animals will mate when the females come in heat (estrus); if they conceive, the pregnancy will usually terminate at about 3 or 4 months. There are reports of rare cases of sheep-goat crosses delivered alive at term.

3.7 Housing

Climatic data can be used as a guide for the type of housing, insulation and ventilation. A housing design should be workable, affordable to goat keepers. A detailed cost analysis is required before starting the construction for housing and shelter for goats for commercial purpose. Housing structure need some provision to rear type of goats such as meat, dairy, and fibre types. Therefore, it is necessary to include farm structure as per requirement. A dairy goat will need provision for milking and milk storage. Fibre producing goats will require an area for shearing, fibre collection, and storage. If the farm is for demonstration and training purpose and necessary facilities need to develop for biosecurity measures and safe visit of visitors. Simple housing for goats is adequate in areas where the weather is mild because goats do best when they are outside on pasture or in an area where they can exercise freely.

Space requirement: The size and natural behaviour of goat must be taken into consideration for finalizing space requirement of goats. The natural movement of goat, bipedal stance, resting, feeding, and socializing should be considered for providing space requirement. Confinement causes seasonal parasitic and coccidiosis problem in goat shed. As a thumb rule, the space requirement will be around 1.4 m^2 (15ft^2) for adult goat. The space requirement is more for concrete space as compared to elevated slatted floors. An additional area of 2.3 m^2 (25 ft^2) of well drained and fenced area is required for exercise. It is necessary to provide adequate space for goats at feeding trough. Therefore, it is necessary to provide adequate feeding trough space to avoid aggression and low growth of some individuals.

Table 3.10 Optimum floor space requirements for different categories of goats

	Floor space (m^2/goat)	
Category/Age of goats	Covered space	Open space
0–3 months	0.20–0.25	0.40.0.50
3–6 months	0.50–0.75	1.00–1.50
6–12 months	0.75–1.00	1.50–2.00
Adult animal	1.50–2.00	3.00
Bucks, pregnant and lactating does	1.50–2.00	3.00–4.00

The floor space for different age groups should be slightly more in hot weather conditions as compared to cool weather. In tropical climates, huddling is disadvantageous for health of the animals and their productivity. The air space requirement per animal is also important as it is necessary to have clean air and the concentration of pollutants is directly proportional to stocking density. The suitable floor space for goat of different age groups is presented in Table 3.10.

A good fencing is particularly important. Fences must be properly constructed with woven wire approximately 6 ft. high. Fencing for goats is different from fencing for cows; a woven wire fence is best for containing goats because they can climb a rail fence. In addition, a special type of bracing at corners of the fence is necessary because goats can walk up a brace pole and jump over the fence. The equipment such as dehorners, a tattoo set, hoof cutter, a grooming brush, an emasculator, and a balling gun for administering boluses (large pills for dosing animals) are required.

3.8 Disease Control

Diseases cause losses due to morbidity, mortality, veterinary cost, and production loss. A wide range of bacterial, viral, parasitic infections cause heavy morbidity and mortality in livestock and cause huge economic loss to farming community. In livestock production system, diseases problems are inevitable; hence, it is required to adopt suitable herd health management and preventive measures to control diseases and to reduce the production losses. The disease management option improves the animal welfare as well as also provides safer food to consumers.

3.8.1 Biosecurity Measures

Biosecurity is a system of practices adopted in the livestock farm that prevent the transmission of infectious diseases into the herd or to premises. Effective biosecurity management practices prevent the spread of disease by minimizing movement of biological organisms and their vectors within the premises. These approaches help in cost effective health management, consumers' confidence on farm products, and better animal welfare issues.

The immunity of animals depends upon host factors (health, immune status, etc.), environmental factors (temperature, stocking rate, pasture condition, etc.), and the disease agent itself. Therefore, it is necessary to follow area specific suitable packages of practices for effective biosecurity measures. The following points need to be followed for adopting biosecurity measures.

1. Animal identification system needs to be implemented in farm/flock.
2. Effective quarantine practices should be implemented in farm. The introduction of outside animals to healthy flock should be done after meticulous clinical examination, vaccination, and deworming.
3. Region specific vaccination programme should be followed.
4. Provide clean area for restraint, treatment, and isolation of sick animals.
5. A regular practical schedule to observe the animal and identify sick animal in flock.
6. Treatments in sick ward and avoid repeated use of single needle for injection.
7. Prompt removal of placentas and aborted foetuses. Placentas and aborted foetuses can harbour thousands of infectious organisms that can spread infections to other animals.
8. Proper personal hygiene of farm workers (Follow schedule).
9. Proper manure management by adopting the practice of compost and vermin-compost preparation.
10. Routinely clean and disinfect feeding equipment, which can be done with chlorine, iodine, or quaternary ammonia products (QAPs).
11. Routinely clean and disinfect equipment used to medicate animals, especially equipment used on multiple animals.
12. Check all possibility of contamination by vehicles, equipment, clothing, and other contaminated inanimate objects.
13. Safe and timely disposal of carcasses of dead animals as per regional guidelines of state government.
14. Proper storage of feed /feeder and prevention of water contamination.
15. Devise the programme to the control of non-livestock vectors threat to disease transmission (birds, rodents, insects, cats, etc.).
16. Transport animals in clean vehicles and always properly vaccinated animals.
17. Visitors should be permitted through antiseptic foot path way.

3.8.2 Goat and Sheep Diseases

Blue tongue: Bluetongue is an infectious, arthropod-borne, viral disease of ruminants caused by an Orbivirus transmitted by Culicoides biting midges, which severely affecting sheep and cattle. Bluetongue infection of goats is typically an inapparent infection. Viruses from the western lineages circulate in Africa, the Caribbean, and the Americas, whereas those from eastern lineages are endemic in Asia, Indonesia, and Australia.

Johne's disease is a serious health problem in goats and sheep. The disease is caused by *Mycobacterium avium* subsp. *paratuberculosis* (MAP). It is one of the most important chronic wasting diseases of adult goat and sheep with significant economic loss. The susceptible animals pick up infection through contaminated food and water. Milk is an important source of transmission of disease in suckling animals. It is most common in goats at 2–3 years of age and the affected goats become unthrifty, emaciated, and unproductive. It is the disease associated with intensive management system. The affected animals' loose body weight continuously and become weak with hide bound condition, staggering gait, rough coat, anaemia, intermandibular oedema, intermittent diarrhoea followed by death. Treatment is neither effective nor economically feasible with antimicrobial and antitubercular drugs. The control of Johne's disease is based identification and elimination of infected animals and prevention of new infections by adopting best management practices and vaccination.

Caseous Lymphadenitis is a disease that occurs more frequently in adult animals. The pathogenic organism grows in lymph gland, and causes a large development of caseous material (a thick, cheese like accumulation) to form nodule.

Caseous lymphadenitis is characterized by nodules in the lymph area (throat). Caseous lymphadenitis is a highly contagious chronic bacterial disease in goats and sheep caused by Corynebacterium pseudotuberculosis. It mainly occurs in the North and South America, Australia, New Zealand, Europe, and South Africa. This disease is clinically characterized by abscess formations in the skin, internal, and external lymph nodes, and internal organs. Goat or sheep can pick up infection from contaminated soil of the premises, water troughs, grazing on infected pasture, direct contact with a ruptured abscess. Ruptured abscess and nasal / oral secretions from animals with pulmonary abscesses are the major source of contamination. Thereafter, the organism multiplies and spread throughout the body via the bloodstream and lymph nodes, the vital organs lungs, kidney, and liver and also exhibit abscesses.

When one of the infected nodes is opened, a thick caseous material is exposed. The disease is difficult to control. Infected animals should be isolated; then infected nodes are opened and flushed with H_2O_2. Lancing abscesses should be performed under the direction of a veterinarian as this disease is transmissible to humans. Treatment with sensitive antibiotics is generally not rewarding, while subcutaneous abscesses can be managed with surgical drainage and antiseptic dressing with iodine solution, with a precaution to dispose the pus to avoid soil contamination.

Contagious ecthyma, also called *sore mouth*, is contagious diseases of sheep and goat and humans. Contagious ecthyma is a viral disease of goats and sheep, which is highly contagious in nature. Contagious ecthyma is caused by an epitheliotropic Para poxvirus that enters the goats through skin abrasions. The virus replicates in proliferating keratinocytes in the damaged epidermis. The distribution of the disease is worldwide. The disease is characterized by pustular, scabby, and proliferative lesions on the muzzle, lips, face, ears, teats or vulva. It causes unthriftiness, painful lesions, associated with economic loss. This disease can occur at any time, but outbreaks are most common in winter season specially grazing on frozen grasses. The disease appears in the form of papules which progress rapidly to vesicles, pustules, and scabs. Crusty proliferative lesions typically form on the lips but can also affect the face, ears, scrotum, teats or vulva. Large masses of granulation tissue develop under the scabs; develop in approximately 6–7 days after initial lesion formation. The skin crusts gradually dried and fell off, leaving areas of alopecia and depigmented skin.

Enterotoxaemia (overeating disease) caused by *Clostridium perfringens* types C and D and cause frequent mortality in goats and sheep. The disease affects lambs in feedlot condition as well as grazing in field condition. The principal cause of caprine enterotoxaemia worldwide is *C. perfringens* type D. Animals of all ages are susceptible. Sudden Change in weather and diet towards more carbohydrate rich feed may lead to the predispose enterotoxaemia. Most outbreaks occur in goats raised under intensive or semi-intensive management conditions. The well-fed kids and feedlot kids on heavy grain ration are found to be more susceptible. Good class of animals affect commonly due to overeating of concentrate diet competitively and grazing on lush pasture. Toxoid vaccination given to kids followed by a second dose in 2 weeks and booster doses each year will control the disease. This disease, sometimes called overeating disease, can be prevented by avoiding feeding excess milk and grain. Vaccinating pregnant ewes can prevent losses from this disease among Young lambs.

Brucellosis: Brucellosis occurs in both sheep is a zoonotic disease as it is transmissible to humans *Brucella melitensis* is the etiological agent of caprine brucellosis. It is an infectious zoonotic disease and has significant economic impact on the livestock industry and public health. The disease is present in 5 out of the 7 continents (South and North America, Europe, Asia, and Africa) and remains a major problem in the Mediterranean region, the Middle East, Central and Southeast Asia, sub-Saharan Africa, and parts of Latin America with the seropositivity of 2–10% in goat. The disease is mainly transmitted through aborted foetal content and vaginal exudates. The major transmission of the disease in the flock occurs due to ingestion of infected colostrum and milk. The infection is transmitted through wound, mammary gland or during breeding time. After infection the organism grows in chorionic epithelium and spread to chorion and uterine mucous membrane. The treatment option is usually not economically feasible or therapeutically effective. No successful treatment can be prescribed, as the organism is an intracellular parasite. However, antibiotics like streptomycin, oxytetracycline or enrofloxacin works well. All the animals should be tested for brucellosis, and reactor animals or suspects must be slaughtered.

Urinary calculi occur in all livestock including sheep and goat. Urinary calculi occur when the salts in the body are precipitated and form stones that may lodge in the kidneys, ureters, bladder, or urethra. Calculi result from the formation of stones in the urinary tract of male goats and sheep, especially wethers due to an incorrect balance of calcium and phosphorus in the diet. Inability to urinate, belly kicking, excessive stretching, and general restlessness are recognizable symptoms. Diets must be carefully balanced and animals should have continuous access to fresh water at all times. It is essential to provide clean water to flock to prevent the calculi.

The ideal ratio of calcium to phosphorus is 1.6:1.0. Ammonium chloride can be included in the ration to help prevent urinary calculi.

Mastitis is an inflammation of the udder and occurs due to not maintaining proper sanitation for udder health and improper milking. Mastitis is a common clinical condition of udder. The disease is characterized by swelling of udder, decreased milk production, discolouration of milk with appearance of pus in late stage and atrophy of udder in neglected cases. It is caused by many bacteria's (*Staphylococcus aureus* and *E. coli, Pseudomonas spp., K. pneumoniae, C. pseudotuberculosis, M. haemolytica*); mycoplasma (Contagious Agalactia, caused by *M. agalactiae* and *M. mycoides var. mycoides*), and fungus.

It is considered the diseases of mismanagement, where unhygienic conditions increased of risk of diseases occurrence. Gangrenous mastitis (Blue bag) is also observed in goats, in which infected udder become cold, blue coloured and may slough off. Treatment is not successful but attempt may be made with antibiotic therapy. The milk becomes curdled and stringy. This should be treated suitably with systemic antibiotic treatment intramuscularly or intravenously or intramammary application. Hygiene, sanitation, cleanliness of farm premises and milk utensils, regular washing of udder and milkers' hands are some important key factors to prevent mastitis in the farm.

Peste des petits ruminants (PPR): Peste des Petits Ruminants (PPR), also called goat plague, is a widespread viral disease caused by a Morbillivirus (*Paramyxoviridae*). The disease is considered to be a major constraint in the development of small ruminant productivity among smallholders, mainly by causing high mortality in young stock [56]. There is a single serotype of PPR virus, but four distinct genetic lineages. Morbidity and mortality are high when occurring in native sheep and goats' populations. The pre-acute and acute forms are seen mainly in goats. This disease is characterized by high fever, severe depression, decreased appetite, watery nasal discharge progressing to profuse muco-purulent discharge occluding the nostrils, respiratory distress, and sign of bronchopneumonia, conjunctivitis, erosive, and necrotic lesions in the mouth (dental pad, hard palate, cheeks), severe diarrhoea, dehydration, and weight loss. Death usually occurs within 1 week of the onset of illness and earlier in per acute cases. At the face of outbreak, diseased animals should immediately be separated from the healthy flock and all unaffected animals should be vaccinated. In established goat farms, introduction of new animals should be done only after PPR vaccination. Transportation stress precipitates PPR in goats and sheep, therefore, due vaccination is advised before transportation. General managemental practices such as quarantine, proper disposal of carcasses and contact fomites, decontamination and restrictions on importation of goats from affected areas should be strictly implemented.

Foot and mouth disease (FMD): FMD is one of the most regulated livestock diseases on the national map for animal diseases in the most of countries. The disease imposes considerable economic losses due to reduction of milk, meat, and heavy kid/lamb mortality. The disease is subclinical in nature in goats and sheep, and some time passed without notice of the farmers. The acute to subacute cases are characterized by vesicular eruptions in the epithelium of buccal cavity, tongue, muzzle, feet, teats, and udder, clinically characterized by lameness, fever and poor appetite. Vesicles formations in foot are very fragile and mostly not observed. When goats become infected with FMDV, the disease may not be diagnosed for a considerable time because symptoms are very mild. In susceptible kids under 1 month of age, morbidity may approach 100 per cent and mortality may be as high as 95 per cent due to myocardial necrosis. In kids, there is focal degeneration of cardiac muscles, which leads to death. The infected animal is the main source of infection. Infection may spread either through direct contact or by contaminated feed and fodder, utensils, and transportation.

Pox: Goat and sheep pox is prevalent in North and Central Africa north of the equator, the Indian subcontinent, the Middle East, China, and Southwest Asia. Pox is a serious disease with high mortality rate of 50% or high. The disease is characterized by skin eruption on whole body. The clinical signs of goat pox virus infection vary with different hosts and in different geographical areas. Goats of all ages are affected but more severe effects are observed in young animals. Goat and sheep pox is an acute to chronic disease characterized by generalized pox lesions throughout the skin and mucous membranes, fever, rhinitis, conjunctivitis, and lymphadenitis. The first signs may include fever, depression, conjunctivitis, lacrimation, and rhinitis. Pox lesions are more easily observed on the hair free parts of the body such as the perineum, inguinal area, scrotum, udder, axilla, and muzzle. Lesions present on the tongue and gums tend to ulcerate. The mortality rates become higher when lesions develop in the respiratory and alimentary tract. Secondary bacterial pneumonia is the common cause of death.

Foot rot occurs when goat and sheep are kept on wet land. Foot rot is one of the most serious and common diseases affecting the sheep industry. It is contagious, but the bacterial organism causing it does not live long in the soil. Hoof trimming of infected feet, followed by treatment with formaldehyde, is the effective treatment of disease. The disease can be treated with systemic medication. The disease can be prevented by regular trimming of hooves and treating the diseased area with a solution of one-part formalin solution to nine parts of water. Formaldehyde must be used with caution because the fumes are damaging to the respiratory system of both the sheep and the person applying the formaldehyde.

Caprine arthritis encephalitis (CAE): This disease affects the central nervous systems of young kids. The arthritis form of the disease occurs in mature goats and affects the feet and other limb joints. Other symptoms are loss of condition and "hard udder" (a firm, swollen udder in freshly kidded goats). There is no effective treatment for the disease. Preventive methods include reducing contact with infected animals (e.g., at shows and sales), blood-testing suspects, and removing infected animals from the herd. The highest prevalence occurs in Canada, France, Norway, Switzerland, and the USA with the sero-prevalence rate exceeded 65% in these countries. In countries actively importing dairy goats, such as Kenya, Mexico, New Zealand, and Peru, the overall prevalence of CAE infection was often less than 10% [51]. All breeds of goats are susceptible to CAE; however, serological surveys indicate that the disease is most common among the dairy goat breed. CAE virus is transmitted naturally in the neonatal period from an infected adult goat to the kid through consumption of infected colostrum and milk. There is evidence to suggest that CAE can also be transmitted directly from goat to goat possibly through saliva, nasal secretions, urine, or faeces.

Caprine pleuro-pneumonia results in high mortality. The services of a veterinarian should be obtained to help control this disease.

Ketosis is a metabolic disease that affects ewes in late pregnancy particularly due to high-energy requirement. The problem is that ewes carrying twins or triplets must break down body fat to provide their energy needs in later pregnancy. It is possible that fat breakdown may not be sufficient for the glucose needs of these ewes, resulting in hypoglycaemia. *Ketosis*, or pregnancy disease, rarely occurs in goats. The affected goats are in pain and cannot walk. The amount of feed should be reduced in the latter part of the pregnancy to avoid fat conditions. The amount of feed should be reduced in the latter part of the pregnancy to avoid fat. The goats have clean, fresh water at all times helps prevent ketosis. Feeding some high-energy grain such as corn, barley, or milo may prevent pregnancy disease.

Gastrointestinal nematodiasis: Among nematodes *Haemonchus contortus* is widely prevalent and most pathogenic in goats and sheep causing severe diarrhoea and anaemia due to their voracious blood sucking nature in abomasum. *Haemonchus contortus* infection leads to gastric haemorrhages and the animals may die suddenly. Similarly the intestine is also affected with Ostertagia, *Trichostronglyus colubriformis* and *Cooperia* spp., *Strongyloides papillosus*, Nematodirus spp., *Bunostomum trigonocephalum,* and *Gaigeria pachyscelis*, associated with acute to moderate green or black coloured foul smelling diarrhoea with continuous soiling of perineum and tail. The infection causes catarrhal or haemorrhagic enteritis, retarded growth of animals and heavy infection may cause death. In such prolong cases, intermandibular oedema (Bottle Jaw) and anaemia may develop. In chronic cases, there is dullness, loss of body condition, gradual weight loss and affecting production and performance.

Coccidiosis: It is an important protozoan infection of intestine particularly in kids, which affect mostly age group of 2–6 month. Several *Eimeria* sp. are responsible for coccidiosis in goats. In coccidiosis, the damage done to the host is essentially the result of intestinal cell destruction occurring when coccidian cysts are released from host cells. In young kids, coccidiosis is clinically manifested by the presence of most conspicuous sign of diarrhoea. Faeces are loose, foul smelling, discoloured with or without blood. Animals may show abdominal pain, anaemia, inappetance, and weight loss. The look of animal is unthrifty with rough coat. Death may occur in severe cases. Clinical disease is common after the stress of weaning, feed change or transportation. Survived kids have poor appetite and poor body weight gain.

The clinical signs may be correlated with oocystic faecal counts. The oocysts may be detected in faecal examination. Coccidiosis is self-limiting, if not severe or acute and clinical signs usually subside spontaneously. Coccidiosis is mainly a managemental disease. It can be controlled by avoiding overcrowding, keeping younger stock separate from the adult, proper cleaning of sheds and replacement of soil at a regular interval of time, sprinkling lime powder inside the sheds, medication may also be done to prevent the oocysts load. Chemotherapeutic agents recommended for treatment and control of coccidiosis in kids.

Ectoparasitic infestation: Goats are subject to many severe external and internal parasites. The external parasites include *lice* and *mites*. The animals may be sprayed, dipped, or dusted with appropriate insecticides similar to those used on dairy cows. Dairy feed stores are usually good sources of materials and advice for treatment. Goats mainly infested with lice (Bovicola caprae, Linognathus vituli), mites (Sarcoptes scabiei, Psoroptes caprae, Chorioptes caprae, and Demodex caprae), and ticks (Ixodes ricinus, *Boophilus microplus, B. annulatus*). These ecto-parasites cause cutaneous itching, irritation, restlessness, dermatitic lesions, anaemia, tick paralysis, and are responsible for transmission of diseases and associated with poor production. Their infestations are common during hot and humid climatic condition. Sarcoptes scabies causes sarcoptic mange in which mites pierce the skin of the non-hairy parts, e.g. head, face, and ears and results dry scabs. They occur more frequently during winter and spring seasons. It becomes chronic, affecting adversely the overall productivity of the goats. Diagnosis is made by lesions and by examined for mites and their eggs, larvae, and nymphs in scrapings from lesions.

Muscles Disease: Sheep are subject to a nutritional disease known as white muscle disease. The disease can be prevented by supplementing selenium during the last one-third of pregnancy and the lambs should be given as feedmix.

Shipping fever is a disease which usually affects lambs after the stress of transportation. Antibiotics and sulfonamides are usually effective as treatment. Care in transporting lambs helps prevent this disease.

Grass tetany (or *grass staggers*) is due to a magnesium deficiency and also more pronounced in lactating ewes when they are put onto lush pasture. An injection of 50–100 ml of 20% calcium borogluconate or an injection of magnesium sulphate should give rapid recovery.

3.9 Energy Management in Goat and Sheep Farming

Animal production is basically an input–output system to which law of conservation of energy applies in the same way as it does for any other energetic system. Energy in output (production, losses) requires an equal amount of energy input. It has been suggested that less fossil fuel will be available for animal production by 2050 and different forms of energy should be explored to be used for animal agriculture. It has been estimated that 35 KJ of fossil energy were used to produce 1 kg of feed lot beef [57].

Sustainability in livestock farming system can be achieved by managing energy. Most of the rural areas of developing countries are lacking conventional type of electric supply system. This is mainly due to inadequate production of hydroelectric generation potential of developing countries. The livestock farming system need continuous energy flow system to bring sustainability in livestock farming in rural areas which are oddly distributed around the peripheral parts of developing countries. On the basis of geographical location, rural areas can generate different sources of non-conventional energy. Renewable energy and livestock farming are complementing with each other to bring sustainability in the system. Wind, solar, and biomass energy can be harvested to provide long-term supply of electricity not only for livestock farming but also for other agricultural activities.

3.10 Value-Added Biopharmaceuticals from Goat Farming

The development of therapeutic proteins technology from microbes to mammary gland bioreactor has proven to be a better option for human protein expression. Interestingly, it has been observed that the goat milk contains several types of human proteins. Thus, especially genetically modified goat can be used for the production of therapeutic recombinant proteins due to the production of adequate milk, at the cost of low budget small ruminant farmhouse with minimum maintenance expenditure. The state of the art of production of transgenic goat has diverted attention of pharmaceutical industrialists for therapeutic proteins production. The genetically engineered goat is supposed to be a better alternate as small ruminant for the use as bioreactor for the production of recombinant therapeutic proteins in their mammary gland. The first human biological drug (antithrombin-AT) was approved by the European Agency of the Evaluation of Medical Products (EAEMP), and after in the USA, by Food and Drug Administration (FDA). The goat produces about 7–8 folds increase in the amount of serum from rabbit. Besides this, the goat serum contains about 20 mg/ml of total IgG, which is about 2–3 times more than rabbit serum. Thus, goat model bioreactor is more cost effective alternate for producing large amount of antibody in a highly commercialized competitive market. The most important fact is getting antibody from single animal, rather than several different individuals.

Presently, the problem related to goat milk marketing is due to inadequate sequential of commercial process technologies involve in logistic process. Generally, the raw milk is thoroughly inspected before its transfer to logistic systems for distribution. So, it is necessary to have additional marketing skill. Generally, kids are raised to different market weights in different market seasons. At present powdered goat milk is gaining commercial important due to its more shelf-life without any substantial changes in quality [58, 59]. In addition, powdered goat milk is easy to handle by logistic system for distribution at the cost of low expenditure [60, 61]. The dehydration technology is helpful for the consumers with year-round marketable product. Spray drying is a common technology for making powered milk from any milk. Powdered caprine milk is used for variety of products like ice cream, chocolate, yogurt, infant formula, cheese [62]. The main reason for increase demand of goat milk is for production of low lactose or lactose free products. This is required for the people having lactose intolerance, and the number of lactose intolerance people is in increasing order. Goat milk infant formula is also known as breast milk formula. It is a formulated product with the combination of goat milk as raw material. This market is to fulfil the requirement of bottle-feeding or cup-feeding babies allergic for cow's milk. Dairy goat co-operative (DGC) ranks the first in terms of revenue share in global market of goat milk infant formula.

References

1. Hole F (1996) The context of caprine domestication in the Zagros region. In: Harris DP (ed) The origins and spread of agriculture. UCL Press Limited, University College London, London, p 594
2. Foote WC (2004) History and accomplishments of the international goat association during the last 20 years. Small Rumin Res 51:115–121
3. Mason IL (1984) Evolution of domesticated Animals. Longman, London, p 452
4. Morand-Fehr P, Boutonnet JP, Devendra C et al (2004) Strategy for goat farming in the 21st century. Small Rumin Res 51:175–183
5. Oklahoma State University (OSU) (2004) Breeds of Livestock. http://www.ansi.okstate.edu/breeds/goats/
6. Rout PK, Saxena VK, Khan BU et al (2000) Characterization of Jamunapari goats in their home tract. Anim Genet Resour Inform 27:43–52
7. Rout PK, Kumar A, Behera BK (2020) Goat production and supply chain management in tropics. CAB International, Wallingford
8. Mason IL (1980) Sheep and goat production in the drought polygon of Northeast Brazil. World Anim Rev 34:23–28
9. Lanari MR, Taddeo H, Domingo E et al (2003) Phenotypic differentiation of exterior traits in local Criollo goat population in Patagonia (Argentina) arch. Tires. Dummerstorf 46(4):347–356
10. Devendra C (2007) Goats: biology, production and development in Asia. Academy of Sciences, Malaysia
11. Mason IL (1996) A world dictionary of livestock breeds, types and varieties, 4th edn. C.A.B. International, Wallingford
12. Neopane SP (2000) Selection for the improvement of the productivity of hill goats in Nepal. In: 7Th international conference on goats, 15-18 May 2000, Tour France, 19-21 May 2000 Poitiers, France, 1. pp 206-208
13. Campbell QP (2003) The origin and description of southern Africa's indigenous goats. S Afr Soc Anim Sci 4:18–22
14. FAO (2004) Production yearbook 2003. Food and Agriculture Publishers, Rome, 57: 213–215, 241–242
15. Alexandre G, Aumont G, Mandonnet N et al (1999) The creole goat of Guadeloupe (French West Indies): an important genetic resource for the humid tropics. Anim Genet Resour Inform 26:45–55
16. Shrestha JNB, Fahmy MH (2007) Breeding goats for meat production: a review. Selection and breeding strategies. Small Rumin Res 67:113–125
17. National Pygmy Goat Association (1932) 149th Ave SE, Snohomish, WA. http://www.npga-pygmy.com
18. Casey NH, Niekerk V, WA (1988) The Boer goat origin, adaptability, performance testing, reproduction and milk production. Small Rumin Res 1: 291–302
19. Mason IL (1981) Breeds. In: Gall C (ed) Goat production, chapter 3. Academic Press, London
20. Cheng P (1984) Livestock breeds of China. Animal production and health paper 46, FAO, Rome
21. Özcan H, Yalçin BC (1977) özel Zootekni (special animal husbandry) (mimeograph). Istanbul üniv. Vet. Fak., Istanbul. Özcan, H., Yalçin, B.C. Evrim, M. and Oğuz, H. 1980. Imrozkoyunirkininyarientansifkosullardakiverim performansi. I.Büyüme, Yasamagücüvecanliağirlik. (production performance of Imroz sheep under semi-intensive conditions. I. Growth, survival rate and body weight). İstanbul Üniversitesi Veteriner Fakültesi dergisi 6(1–2), 1–9
22. Gurung NK, Solaiman SG (2010) Goat breeds. In: Solaiman SG (ed) Goat science and production. Wiley- Blackwell Publishers, New York, pp 339–357
23. Wright S (1921) Systems of mating I. The biometric relations between parent and offspring. Genetics 6:111–123

24. Lush JL (1937) Animal breeding plans, 1st edn. Collegiate Press, Ames
25. Hazel LN (1943) The genetic basis for constructing selection index. Genetics 28:476–490
26. Galal S (2005) Biodiversity in goats. Small Rumin Res 60(1):75–81
27. Safari E, Fogarty NM, Gilmour AR (2005) A review of genetic parameter estimates for wool, growth, meat and reproduction traits in sheep. Livest Prod Sci 92:271–289
28. Fogarty NM (1995) Genetic parameters for live weight, fat and muscle measurements, wool production and reproduction in sheep: a review. Anim Breed Abst 63:101–143
29. Prakash V, Prince LLL, Gowane GR et al (2012) The estimation of (co)variance components and genetic parameters for growth traits and Kleiber ratios in Malpura sheep of India. Small Rumin Res 108:54–58
30. Jawasreh K, Ismail ZB, Iya F et al (2018) Genetic parameter estimation for pre-weaning growth traits in Jordan Awassi sheep. Veterinary World 11(2):254–258
31. Singh D, Kumar R, Pander BL (2006) Genetic parameters of growth traits in crossbred sheep. Asian Australas J Anim Sci 19(10):1390–1393
32. Lôbo AMBO, Lôbo RNB, Paiva SR et al (2009) Genetic parameters for growth, reproductive and maternal traits in a multibreed meat sheep population. Genet Mol Biol 32(4):761–770
33. Notter DR (1998) Genetic parameters for growth traits in Suffolk and Polypay sheep. Livest Prod Sci 55(3):205–213
34. Lavvaf A, Noshary A (2008) Estimation of genetic parameters and environmental factors on early growth traits for Lori breed sheep using single trait animal model. Pak J Biol Sci 11 (1):74–79
35. Pickering NK, Dodds KG, Blair HT (2012) Genetic parameters for production traits in New Zealand dual-purpose sheep, with an emphasis on daggines. J Anim Sci 90:1411–1420
36. Ghafouri-Kesbi F, Baneh H (2012) Genetic parameters for direct and maternal effects on growth traits of sheep. Arch Tierzucht 55(6):603–611
37. Bahreini Behzadi MR, Shahroudi FE, Van Vleck LD (2007) Estimates of genetic parameters for growth traits in Kermani sheep. J Anim Breed Genet 124:296–301
38. Jafari S, Razzagzadeh S, Hashemi A, et al (2011) Estimation of genetic parameters and genetic trend for body measurements (biometric characteristic) and their correlations with yearling live weight in Makuie sheep. In: 1st national agriculture congress. pp 27–30
39. Illa SK, Gollamoori G, Nath S (2019) Direct and maternal variance components and genetic parameters for average daily body weight gain and Kleiber ratios in Nellore sheep. Trop Anim Health Prod 51:155–163
40. Gowane GR, Chopra A, Prince LLL et al (2010) Scope of indirect selection for wool traits in Bharat merino sheep in semi arid region of Rajasthan. J Appl Anim Res 37(1):97–100
41. Puntila ML (2007) Genetic parameters for wool traits in Finn sheep lamb. Agric Food Sci 16:124–135
42. Mandal A, Neser FWC, Roy R (2009) Estimation of (co)variance components and genetic parameters of greasy fleece weights in Muzaffarnagri sheep. J Anim Breed Genet 126:22–29
43. Murphy TW, Berger YM, Holman PW (2017) Estimates of genetic parameters, genetic trends, and inbreeding in a crossbred dairy sheep research flock in the United States. J Anim Sci 95:4300–4309
44. Ali Saghi D, Shahdadi AR (2017) Estimates of genetic and phenotypic parameters for reproductive traits in Iranian native Kordi sheep. Acta Scientiar Anim Sci 39(3):323–328
45. Brown DJ, Fogarty NM, Iker CL (2015) Genetic evaluation of maternal behaviour and temperament in Australian sheep. Anim Prod Sci 56(4):767
46. Matheson SM, Bünger L, Dwyer CM (2012) Genetic parameters for fitness and neonatal behavior traits in sheep. Behav Genet 42:899–911
47. Tortereau F, Doucet M, Gauthier P (2018) Genetic parameters for lamb vigour and early mortality in two French sheep populations. Proceedings of the world congress on genetics applied to livestock production 11.182

48. Smith MC (1980) In: Morrow DA (ed) Caprine reproduction. Current therapy in theriogeniology diagnosis, treatment and prevention of reproductive diseases in animal. W.B. Saunders Company, Philadelphia
49. Gall C (1981) Milk production. In: Gall C (ed) Goat production. Academic Press, London, pp 309–344
50. Devendra C, Burns M (1983) Goat production in the tropics. Commonwealth Agricultural Bureau International Wallingford
51. Greyling JPC (1988) Certain aspects of reproduction physiology in the boar goat doe. Ph D, University of Stellenbosch, South Africa
52. Jainudeen MR, Wahid H, Hafez ESE (2000) Sheep and goats. Reproduction in farm animals (7th edn)
53. Sheep Industry Council (2004) Sheep production handbook. Sheep Industry Development Program, Denver
54. NRC (1985) Nutrient requirements of sheep, 6th revised edn. National Research Council, National Academic Press, Washington
55. NRC (2007) Nutrient requirements of small ruminants. National Research Council, National Academic Press, Washington
56. Smith MC, Sherman DM (2009) Goat medicine. Wiley-Blackwell
57. Hillel D, Rosenzweig C (2008) In: Chavan E, Bernstein A (eds) Biodiversity and food production. In sustaining life: how human health depends on biodiversity. Oxford University Press, Oxford, pp 325–381
58. Silanikove N, Leitner G, Merin C et al (2010) Recent advances in exploiting goat's milk: quality, safety and production aspects. Small Rumin Res 89(2):110–124
59. Fonseca R, Helena GB, Ana Cláudia D et al (2013) Juliomys rimofrons Oliveira and Bonvicino, 2002 (Rodentia:Cricetidae): distribution extension. Check List 9(3):684–685
60. Bylund G (2003) Dairy processing handbook. Tetra Pak Processing Systems ABLund
61. Tehrany E, Sonneveld K (2010) Packaging and the self-life of milk powders. Food packaging and practical guide. CRC Press, Boca Raton, pp 127–141
62. Kajal MFI, Wadud A, Islam MN et al (2012) Evaluation of some chemical parameters of powder milk available in Mymensingh town. J Bang Agric Univ 10(1):95–100

Cattle and Buffaloes Farming

4

4.1 Introduction

Cattle were domesticated in Asia and Europe during New Stone Age and domestication occurred 10,000 years ago. Humped cattle were found in tropical countries and *Bos taurus* cattle were observed in temperate region. Cattle along with buffaloes provide food, fibre, fuel, and draft power to our world. Milk and draft power are the primary products all over the world; however, in some developed countries beef also plays a major role in cattle industry. The domestication of buffaloes occurred around 5000 years ago. Asian buffalo includes two subspecies known as the River and Swamp types. River buffaloes are generally large in size, with curled horns and are mainly found in India, Pakistan, and in some countries of western Asia. The river buffaloes are primarily used for milk production and also for meat and draught purposes. Swamp buffaloes are stocky animals with marshy land habitats and are primarily used for draught power in paddy fields and are also used for meat and milk production. Swamp buffaloes are mostly found in South East Asian countries. A few animals can also be found in the north eastern states of India [1]. Buffaloes are known to be better at converting poor-quality roughage into milk and meat. They are reported to have a 5% higher digestibility of crude fibre than high-yielding cows; and a 4–5% higher efficiency of utilization of metabolic energy for milk production [2].

4.2 Cattle Breeds

Cattle are the major livestock species and different cattle breeds are recognized throughout the world. A breed is defined as a race or variety, the members of which are related by descent and similar in criteria distinguishable characteristics. Breeds are formed through genetic isolation and either natural adaptation to the environment or selective breeding, or a combination of the two. Herd books and breed registry associations were established to maintain the "purity" of each breed and to improve each breed. Purebred refers to purity of ancestry as established by pedigree record of

P. K. Rout, B. K. Behera, *Sustainability in Ruminant Livestock*,
https://doi.org/10.1007/978-981-33-4343-6_4

each individual. Purebreds are the individuals those have pedigrees recorded in their respective breed registry associations or with farmers.

4.2.1 Dairy Breeds

4.2.1.1 Jersey

The breed is native to Jersey Island, U.K. and small size dairy cattle. The body colour is reddish fawn with a compact and angular body. The average milk yield is 4500 kg per lactation.

4.2.1.2 Holstein Friesian

The breed is mainly distributed in northern parts of Netherlands and especially in the province of Friesland. It is a large size dairy breed with well-built body and possesses large udder. The body has typical marking of black and white shade. The average milk production of cow is 6000 to 7000 kg per lactation.

4.2.1.3 Brown Swiss

The mountainous region of Switzerland is the native place of Brown Swiss breed. The average milk yield is 5000 kg per lactation.

4.2.1.4 Red Dane

The breed is mainly found in Denmark. Body colour of this Danish breed is red, reddish brown, or even dark brown. It is a heavy breed. The lactation yield of Red Dane cattle varies from 3000 to 4000 kgs.

4.2.1.5 Ayrshire

The breed is found in Ayrshire in Scotland and considered as most beautiful dairy breed. The breed was also known as Dunlop cattle or Cunningham cattle.

4.2.1.6 Guernsey

The breed has originated from Small island of Guernsey in France. The body colour varies from cherry red to brown. Mahogany and white are a variation in colour. The milk has a golden colour due to an exceptionally high content of beta carotene. Guernsey cows produce around 6000 kg per lactation.

4.2.1.7 Gir

The breed is otherwise known as Bhadawari, Desan, Gujarati, Kathiawari, Sorthi, and Surati. The main habitat is Gir forests of South Kathiawar in Gujarat and also found in Maharashtra and adjacent Rajasthan. The skin colour is white with dark red or chocolate-brown patches or sometimes black or purely red. Horns are peculiarly curved, giving a "half-moon" appearance. Milk yield ranges from 1200–1800 kgs per lactation.

4.2.1.8 Red Sindhi

The breed is also known as Red Karachi and Sindhi and Mahi. The place of origin is Karachi and Hyderabad (Pakistan) and also found in border regions. The body colour is red with shades varying from dark red to light strips of white. Milk yield ranges from 1250 to 1800 kgs per lactation.

4.2.1.9 Sahiwal

The breed has originated in Montgomery region of India and Pakistan. The breed is also locally known as Lola (loose skin), Lambi Bar, Montgomery, Multani, Teli. The body is highly symmetrical and colour is reddish dun or pale red and white patches. The average milk yield of this breed is between 1400 and 2500 kg per lactation.

4.2.2 Draught Breeds

4.2.2.1 Hallikar

The place of origin is the former princely state of Vijayanagaram (Karnataka, India). The body colour is grey or dark grey. The animals are medium in size with prominent forehead, long horns, and strong compact muscular legs. The breed is best known for its draught capacity and especially for its trotting ability.

4.2.2.2 Amritmahal

The breed has originated in Hassan, Chikmagalur, and Chitradurga districts of Karnataka. The body colour is grey and varies from almost white to near black. The muzzle and feat are usually black. Horns are long and end in sharp black points.

4.2.2.3 Khillari

The breed has originated from Sholapur and Sitapur districts of Maharashtra. The body colour is grey-white with long horns turn. The horns are generally black, sometimes pinkish. Bullocks are fast and powerful.

4.2.2.4 Kangayam

The breed is also known Askongu and Konganad. The natural habitat is in Kangayam, Dharapuram, Perundurai, Erode, Bhavani, and part of Gobichettipalayam taluk of Erode and Coimbatore district. Bulls are grey with dark colour hump. Cows are grey or white. The eyes are dark and prominent with black rings around them. The horns are straight and spread apart.

4.2.2.5 Bargur

The breed is distributed around Bargur hills in Bhavani taluk of Erode district in Tamil Nadu. The body colour is brown with white markings. This is a medium size breed with well-built compact muscular body. Known for their speed and endurance in trotting.

4.2.2.6 Umblachery

The breed is otherwise called as Jathimadu, Mottaimadu, Molaimadu, Therkathimadu. The breed is distributed in Thanjavur, Thiruvarur, and Nagapattinam districts of Tamil Nadu. The breed is suitable for wet ploughing and known for their strength and sturdiness. Umblachery calves are red or brown at birth with white marking on the face, limbs, and tail. The legs have white markings below the hocks. The practice of dehorning of bullocks is peculiar in Umblachery cattle.

4.2.2.7 Pulikulam

This breed is distributed in cumbum valley of Madurai district in Tamil Nadu. The breed is locally known as Jallikattumadu, kidaimadu, and sentharai. This is a small size breed and body colour is usually grey or dark grey with well-developed hump. The characteristic feature of this breed is the presence of reddish or brownish spots in muzzle and eyes. Breeds are active, useful draught animals but not fast trotter.

4.2.2.8 Alambadi

The breed is found in Alambadi of Dharmapuri district in Tamilnadu. The body colour is grey or dark grey and white marking observed in forehead, limb, and tail. Horns are backward. It is useful in ploughing.

4.2.3 Dual-Purpose Breeds

4.2.3.1 Tharparkar

The breed is distributed in Tharparkar district of Pakistan and in bordering region of India and Pakistan and also found in Rajasthan. The breed is also known as White Sindhi, Gray Sindhi, and Thari. This is a medium-sized breed with lyre-shaped horn. Body colour is white or light grey. The milk yield varies from 1800 to 2600 kg per lactation.

4.2.3.2 Hariana

The breed is mainly found in Rohtak, Hisar, Jind, and Gurgaon districts of Haryana and also found in Punjab, Uttar Pradesh of India. Horns are small. Hariana cows produce 600 to 800 kg of milk per lactation.

4.2.3.3 Kankrej

The breed is distributed in Southeast Rann of Kutch of Gujarat and adjoining areas of Rajasthan (Barmer and Jodhpur district). The horns are lyre-shaped. The body colour varies from silver-grey to iron-grey or steel black. The gait of Kankrej is peculiar called as 1 1/4 paces (sawaichal). Kankrej is valued as fast and powerful draught cattle. The cows are good milkers, yielding about 1400 kg per lactation.

4.2.3.4 Ongole

The breed is found in Ongole taluk in Guntur district of Andhra Pradesh. The breed has muscular body with well-developed hump. The body colour is white or light grey. The average milk yield is 1000 kg per lactation.

4.2.3.5 Krishna Valley

The breed is distributed in the valley of river Krishna in Karnataka and also found in border districts of Maharashtra. This is a large size breed with deep body frame. The body colour is grey-white with a darker shade on fore quarters and hind quarters in male. Adult females are more whitish in appearance. The average yield is about 900 kg per lactation.

4.2.3.6 Deoni

The breed is locally known as Dongerpati, Dongari, Wannera, Waghyd, Balankya, and Shevera. The breed has originated in Western Andhra Pradesh and also found in Marathwada region of Maharashtra state and adjoining part of Karnataka. The body colour is usually spotted black and white. The milk yield varies from 636 to 1230 kg per lactation.

4.3 Buffalo Production

Asian buffalo or Water buffalo is classified under the genus *Bubalus*, species *bubalis*. The *Bubalus bubalis* belongs to the class Mammalia, subclass Ungulata, order Artiodactyla, suborder Ruminantia, family Bovidae, subfamily Bovinae, tribe Bovini, which includes the following three groups: Bovina (cattle), Bubalina, and Syncerina. Syncerina includes only species *Syncerus caffer* (the African buffalo). Bubalina (the Asian buffalo) includes three species such as *Bubalus depressicornis* or Anoa which lives in Indonesia, *Bubalus mindorensis* which lives in the Philippines, and *Bubalus bubalis* deriving from the domestication of the *Bubalus arnee*, the Indian wild buffalo. Asian buffalo includes two subspecies known as the River and Swamp types. The River buffalo has 50 chromosomes of which five pairs are submetacentric, while 20 are acrocentric. The Swamp buffalo has 48 chromosomes, of which 19 pairs are metacentric. The difference in the diploid number is only apparent. In fact, the large Swamp buffalo chromosome 1 originated from tandem fusion translocation between the River buffalo chromosome 4 (telomeres of p-arm) and 9 (centromere) [3].

4.3.1 Buffalo Breeds

4.3.1.1 Banni

The breed is distributed in Kutch region of Gujarat. The body has deep body conformation with prominent black colour. The ear is short and thin with short and curly horns. The tail is long and continues till fetlock joint. The udder is pendulous with long teats, and thick and prominent milk vein. The breed is famous for their high milk yield and survives in the harsh climate of desert. The average age at first calving is 1250 days. The lactation milk yield, lactation length, and fat% are 2857 kg, 301 days, and 7%, respectively.

4.3.1.2 Bhadawari

The breed is distributed in Bah tehsil of Agra, Chakarnagar, Barhpura blocks of Etawah, Etah, and Jhansi districts of Uttar Pradesh and also Ambah and Porsa tehsil of Morena districts in Madhya Pradesh. This is a medium size breed with wedge shaped body. The eyelids are copper colour and two white lines—chevron at lower side of neck. Forehead—little broad and shallow in middle. Breed is famous for highest fat content in its milk. The average age at first calving is 1340 days. The lactation milk yield, lactation length, and fat% are 1295 kg, 290 days, and 7.8%, respectively.

4.3.1.3 Chilika

The breed is distributed in Chilika lake region in Puri, Ganjam, and Khurda districts in Odisha. Horns are strong and sickle shaped in adult animals. Coat colour ranges from brownish black to black with medium-sized tail. Chilika buffalo grazes throughout the night in marshy land of the lake, feeds on submerged weeds and aquatic vegetation in the lake. In the morning buffaloes are habituated to come to owner's home for drinking fresh water. The average age at first calving is 1362 days. The lactation milk yield, lactation length, and fat% are 500 kg, 255 days, and 8.6%, respectively.

4.3.1.4 Jaffarabadi

The breed is distributed in Junagarh, Bhavnagar, and Amerli districts in Saurashtra region of Gujarat. It is the heaviest of all Indian buffaloes. The horn is long and forms a ring like structure at the tip in a characteristic fashion specific to the breed. The prominent colour is black and tail switch is also black. The average age at first calving is 1356 days. The lactation milk yield, lactation length, and fat% are 2240 kg, 305 days, and 7.7%, respectively.

4.3.1.5 Kalahandi

The breed is native to hilly areas Kalahandi and Nuapada district of Odisha. This is a medium size breed with compact body and strong legs. The milk veins are prominent with medium naval flap. The muzzle is prominent and black.

4.3.1.6 Marathwadi

The breed is distributed in Parbhani, Nanded, Beed, Latur districts of Maharashtra. The coat colour varies from greyish black to jet black and white markings on forehead. The head is prominent with medium length horns. The ears are long and well-developed udder with medium-sized funnel shaped teats. Marathwadi animals are well adapted to hot and humid conditions. The lactation milk yield, lactation length, and fat% are 1120 kg, 318 days, and 8.8%, respectively.

4.3.1.7 Mehsana

The breed is distributed in Banaskantha, Sabarkantha, Gandhinagar, and Ahmedabad districts of Gujarat. This is a medium size breed and coat colour is black or brown and white mark on forehead. The udder is voluminous with well-developed teats. This is a good milk producing breed with higher breeding efficiency. The average age at first calving is 1285 days. The lactation milk yield, lactation length, and fat% are 1990 kg, 317 days, and 6.8%, respectively.

4.3.1.8 Murrah

Murrah is one of the best dairy breeds of buffalo in the world. The breed is distributed in Rohtak, Hisar, and Jind. Karnal districts of Haryana, and Nabha, Patiala districts in Punjab. These animals have massive frame, attractive appearance, and deep body confirmation. The coat colour is typical jet black or glistening black. The udder is capacious with long teats. The horns are curled taking two and half turns inside. The population of this breed is maximum among all the buffalo breeds found in India and used as an improver breed in the entire country. The breed has been taken to several European countries for production of Mozzarella cheese. The average age at first calving is 1550 days. The lactation milk yield, lactation length, and fat% are 1890 kg, 3 13 days, and 7%, respectively.

4.3.1.9 Nagpuri

The breed is distributed in Nagpur, Akola, Wardha, Amrawati, Yavatmal, and Buldhana districts of Maharashtra. This breed is also known as Berari, Ellichpuri. Gaulani, Gauli, or Varadi. The coat colour is black with white patches on face. The udder is moderate in size with conical teats. The horns are long, flat, curved and carried back on side of the neck near to shoulders. Males are good for transport under hot climatic condition and females are good milkers. Low maintenance cost, high feed efficiency, and high milk fat content are the advantageous characteristics of this breed. The average age at first calving is 1710 days. The lactation milk yield, lactation length, and fat% are 1040 kg, 303 days, and 8.3% respectively.

4.3.1.10 Nili Ravi

The breed is distributed in Gurdaspur, Amritsar, Ferozepur, and Muktsar districts of Punjab. Nili Ravi is a medium-sized breed with deep and low set frame body conformation. The coat colour is black or brown, white markings on forehead muzzle/chin. The udder is well developed and capacious. The productivity of this breed is also good and due to its typical markings, it is also known as "Panch-Kalyani". The average age at first calving is 1430 days. The lactation milk yield, lactation length, and fat% are 1945 kg, 306 days, and 6.90%, respectively.

4.3.1.11 Pandharpuri

The breed is distributed in Southern Maharashtra covering Pandharpur. North and South Sholapur districts, parts of Kolhapur, and Sangli districts along with adjoining parts of Karnataka. This is a medium size breed with black coat colour. The horns are long and extending beyond shoulder blade. The udder is predominantly of trough shape with cylindrical teats. This breed has special characteristics of letting down milk at any place and at any time. Milk let down is not a problem in this breed. This type of behaviour has not been recorded in any other breed of buffaloes. The average age at first calving is 1315 days. The lactation milk yield, lactation length, and fat% are 1790 kg, 305 days, and 7%, respectively.

4.3.1.12 Surti

The breed is distributed in Surat, Khaira, Bharuch, and Vadodara districts of Gujarat and adjoining areas of Maharashtra. The coat colour varies from rusty brown to silver-grey. The forehead is broad with sickle-shaped horns. The udder is spacious with medium sized and squarely placed teats. The males of Surti animals are useful for road transport. The average age at first calving is 1350 days. The lactation milk yield, lactation length, and fat% are 1670 kg, 292 days, and 7%, respectively.

4.3.1.13 Toda

The breed is distributed in small area of Nilgiri district of Tamil Nadu. Toda animals are medium sized, powerful, furious, dangerously looking animals. The body is long with broad and deep chest. The coat colour varies from grey to cream. Animals are very strong and live in hamlet and are known to take fights. Long horns growing outward, upward then inward at tips giving semicircle shape. The average age at first calving is 1440 days. The lactation milk yield, lactation length, and fat% are 500 kg, 199 days, and 8.2%, respectively.

4.3.1.14 Kundi

This is a draught breed in the Indus valley Sindh region and is the second most important breed in Pakistan. The coat colour is black with short horns. The body weight of male and females is 700 kg and 600 kg, respectively. Buffaloes are traditionally managed under domestic conditions together with the calf. They are hand milked twice a day. They are mated mainly through natural mating. Some villages also provide artificial insemination. The lactation milk yield during 320 days is 2000 kg, milk fat 7.0%, milk protein 6.0% [4, 5].

4.3.1.15 Lime

The pure Lime breed is believed to have originated from the wild Arna and has been distributed in the high hills, river valleys, and mountains of the Nepal. The coat colour is light brown colour and small body size. The characteristic chevrons of grey or white hair below the jaws and around the brisket, small sickle-shaped horns, curved towards the neck. The body weight is 399 kg. The breed is mainly raised under migratory conditions or semi-stall systems. The breed is a voracious eater and is fed only low-quality feedstuff such as rice, wheat, and millet straw. Small farmers exchange breeding animals within and between villages. Among the migratory herds, male and females are grazed together and mate freely during the breeding season from June to November. The lactation milk yield during 351 days is 875 kg, milk fat 7.0%.

4.3.1.16 Mediterranean Buffalo

The Mediterranean buffalo originates from the Indian buffalo. The breed is distributed in Italy, Romania, and Brazil. It was introduced into Europe with the advent of Islam and the Arab occupation as well as through other central European conquerors in the sixth and seventh Centuries. The coat colour is black predominantly, however brown and dark grey colours are found. Horns are flat at the bottom, backwards and slightly outwards pointed. The body is deep and wide chest as well as a developed pectoral. The back is short. The rump is short. The udder is medium size with squarely placed quarters and halves; the teats are cylindrical. In Italy, the animals are housed loose in paddocks all year long. Average daily milk yield reveals a huge variability, mainly depending on the feeding system. It can range from 3 to 4 kg milk/day for poorly fed animals to 15 kg/day in intensive management systems. In Bulgaria, Romania, TFYR Macedonia, Greece, and Albania, extensive management systems are employed. The lactation milk yield of 270 days was 900–4000 kg, milk fat 8.0%, milk protein 4.2–4.6%.

4.3.1.17 Anatolian

The Anatolian buffalo has been raised in Turkey for centuries, originating from Indian migration (Seventh Century), together with the expansion of Islam. The coat colour is black with long hair and varied tail length. The body weight of animal is 200–500 kg. The breed is distributed in the Black Sea region, North of Middle Anatolia, Thrace, Hatay, Mus, Kars, Dyarbakir, Afyon, and Sivas. The lactation milk yield during 220–270 days is 700–1000 kg, milk fat 6.6–8.1%, milk protein 4.2–4.6%. Products: a semi-hard cheese called "peyazpeyneri" is made from buffalo milk. Ayran is a drink with water and buffalo yoghurt.

4.3.1.18 Azeri or Caucasian

This breed originates from the Indo valley (Indian buffalo). There is some evidence that buffalo were raised in Lorestan (Iran) in the ninth Century B.C. since six engraved buffalo heads have been found on a bronze stick from this period. The coat colour is black with short horns growing backwards. The body weight is 400–600 kg. The breed is distributed in Iran, West Azerbaijan, East Azerbaijan, and the Caspian Sea. Average slaughter weight is 300 kg, at the age of 15 months. Carcass yield is 50 percent. Overall growth rate is 420 g/day. The lactation milk yield during 200–220 days is 1200–1300 kg, milk fat 6.6% [6–13].

4.3.1.19 Bulgarian Murrah

Murrah buffaloes from India were introduced in Bulgaria and local population was upgraded to a new population. Black or black and brown or dark grey in colour. The body weight of adult male and female is 700 kg and 600 kg, respectively. The breed is distributed in Bulgaria, Romania, and South America. Husbandry: Buffaloes are traditionally managed under domestic conditions together with the calf [4, 14, 15].

4.3.1.20 Egyptian

Buffaloes were introduced into Egypt from India, Iran, and Iraq approximately during the middle of the seventh century. The colour is blackish grey and horn shape varies from lyre to sword-shaped. The head is long and narrow; the jaws are long and strong. Ears are long and dropping. The neck is rather long, thin, and straight. The forelegs are rather short and heavy boned. Ribs are wide, deep, and well sprung. The rump is sloping and the tail setting is low. The body weight of adult male and female is 600 kg and 500 kg, respectively. The breed is distributed in peri-urban areas and the Nile delta. The lactation milk yield during 210–280 days is 1200–2100 kg, milk fat 6.5–7.0% [16–19].

4.3.1.21 Kuhzestani or Iraqi Buffalo

This is the biggest buffalo breed in the world. The horns are short and grow upward forming a ring at the end. The body weight of adult male and female is 800 kg and 600 kg, respectively. The breed is distributed in Iran and Iraq. Husbandry: Buffaloes are raised outdoors all through the year. They are housed in paddocks made of local plants (reeds, brushes, palm leaves) with a wall on one side, and three open sides. Average slaughter weight is 400 kg, at the age of 12 months. Carcass yield is 50 percent. The lactation period is 200–270 days and lactation yield was 1300–1400 kg of milk with 6.6% milk fat.

4.4 Genetic Improvement

4.4.1 Selection for Improving Growth and Milk Production

The animals with better structure and function is necessary in a herd for producing seed-stock. Moreover, structural correctness is the major criteria used to assess general appearance for selection purpose and show-ring score-card. Production traits are directly dependent on structural soundness of animal. The farmers should be familiar with anatomical knowledge to identify best animals for higher production.

The animals are selected based on the four major classification traits such as general appearance, dairy character, body capacity, and mammary system. The dairy type cows should have good stature, long neck, prominent milk vein, and strong feet and legs.

The major criteria for each category for cows and bulls are shown in Table 4.1. The final score represents the degree of physical perfection of any given animal. The animals are classified based on score and graded as follows: Excellent (EX): 90–100, Very good (VG): 85–89, Good plus (G+): 80–84, Good (G):75–79, Fair (F): 65–74, Poor (P): 50–64.

Breeders also use linear classification system as an evaluation tool to increase selection accuracy for cows to be productive for longer duration. The linear classification scoring system is used to assess selected traits (Table 4.2). This system would be a useful guide to identify the bulls whose daughters were best fulfilling the farmers' needs. Type classification is an evaluation of body conformation as defined

Table 4.1 Classification traits used for final scores in Holstein cattle

Traits	Emphasis	
	Cows	Bulls
Frame	15	30
Dairy character	20	25
Body' capacity	10	20
Feet and legs	15	25
Udder	40	–

Source: Holstein Association, linear classification programme

Table 4.2 Linear classification scoring system to assess type for selected traits

Trait class	Trait	Description	Score
Form	Stature	Extremely tall	45
		Intermediate	25
		Extremely short	5
Form	Body depth	Extremely deep	45
		Intermediate	25
		Extremely shallow	5
Rump	Rump angle/width	Extremely sloped/extremely wide	45
		Slight slope/intermediate	25
		High pins/extremely narrow	5
Legs and feet	Rear legs (side view)	Extremely sickled	45
		Intermediate	25
		Extremely posty	5
Udder	Fore udder attachment	Fore udder attachment	45
		Intermediate	25
		Extremely loose	5
Teats	Front teat placement	Extremely close	45
		Centrally placed	25
		Extremely wide	5

Source: Adapted from Linear Classification System, Holstein association, USA, 1999

by breed associations. Each breed should develop its true type cow and bull models, which are standards that ensure uniformity and accuracy in type classification. The dairy farmer uses type classification as a management tool in breeding and selection decisions. Records on type traits may be used to develop cow indexes. However, there is a need to harmonize type classification for all breeds.

The performance recording should be carried out effectively to bring the improvement in the desired traits. Identification of animal is necessary before carrying out performance recording. The body conformation should be considered as first parameter and animals should be free from genetic defects. The identification number can be given by ear tags or other effective technique including electronic devices. The recording for body weight should be carried out at different ages. Milk recording traits such as milk yield during different stages of lactation and composition traits should be carried out. The recording of calving information should be taken up such as calving interval, weaning weight, and daily weights. Carcass evaluation and fertility parameters such as scrotal circumference in males and age at first heat or first calving in females are also recorded.

The economically important traits for growth of cattle and buffaloes are weaning weight, post-weaning growth, feed efficiency, carcass merit, and longevity (functional traits). Reproductive performance is considered as most important parameter of flock. Reproductive performance of flock or individual can be improved by modifying the environment with adequate nutrition and health care. The bulls should be selected based on reproductive soundness and conformation traits. Generally, beef producer desires that a cow should calve in every 365 days, and this can be achieved by better managemental condition. The heritability of fertility rate is low (less than 20%), therefore genetic progress will be slowed through direct selection. The heritability for birth weight and scrotal circumference are high and can be achieved by selection.

Weaning weight reflects the mothering ability and milk production capacity of the cow. Weaning weight is commonly expressed as the adjusted 205-day weight, where the weaning weight is adjusted for the age of the calf and age of the dam. Weaning weights of calves are usually compared as a ratio expressed by dividing the calf's adjusted weight by the average weight of the other calves in the contemporary group. Ratios can be used primarily for selecting cattle within the same environmental conditions. Similarly, post-weaning growth measures the growth from weaning to a finished weight. Weaning weight and post-weaning gain are usually combined into a single trait; namely, adjusted 365-day weight (or yearling weight). It is computed as.

Adjusted 365-day weight = (160 × average daily gain) + adjusted 205-day weight.

Average daily gain and adjusted 365-day weight have high heritabilities (40%); therefore, genetic improvement can be quite rapid when selection is practiced on post-weaning growth yearling weight. The heritability of feed efficiency is high (45%), therefore selection can be more effective. Moreover, the genetic correlation between gain and efficiency can be utilized where possible.

Bulls are usually kept in a herd for 3–5 years. Highly productive cows are culled from the herd at an early age due to problems such as skeletal unsoundness, poor udders, eye problems, and lost or worn teeth. However, the highly productive cows are retained in the herd at the age 15 years or older. Body conformation and anatomical structural soundness of an individual are effective parameters for the longevity of animal. There should not be any type of genetic defects or hereditary defects observed in the body.

Selection and culling should be practiced rigorously to achieve breeding goal. A minimum of 500 doses of each young bull's semen should be used in 20 to 30 herds in the progeny test scheme. This distribution process should result in at least 20 daughters from each bull. However, it is not possible to obtain this number of daughters under smallholder farming systems and alternative procedures need to be developed and standardized.

Phenotype is the appearance or performance of an animal determined by the *genotype* (genetic makeup) and the environment in which the animal was raised.

Genotype is determined by two factors: (1) breeding value (what genes are present) and (2) non-additive value (how genes are combined). Environmental factors influence the phenotype through known and unknown effects. Age of dam is a known environmental effect adjusted for 205-day weight (phenotype). The other environmental effects are things like injury or health problems for which it is difficult to adjust the phenotypic record.

Sire selection is first step to bring desired genetic improvement in a herd over the years. Record keeping is required for effective implementation of selection programme. Purebred breeders should provide accurate performance data on their bulls. The trait ratio is useful and comparative if the bulls have been fed and managed in similar environments. The basis for selection and ranking of bulls with respect to their genetic value for different attributes is the expected breeding value (EBV). The EBV for a characteristic such as milk production of daughters is Heritability x Phenotypic Superiority. The latter is the difference between the value for a bull and the mean value for the population in a country or specified area [20].

The analytical methods have been developed that utilize performance records on the individual calf and its relatives. This information is used to calculate breeding values. Breeding values are most frequently reported on birth weight, maternal weaning weight, weaning weight, and yearling weight by using expected progeny differences (EPDs). The heritability of traits is indicative of the progress that can be made by selection. Heritability broadly defines the percentage of total variation that is due to genetic effects. The higher the non-additive genetic variation within a line or breed, the greater is the heterosis in the cross. Usually the breeder chooses the animals to be parents of the next generations, while disposing of others. The selection process may consider a number of suitable traits simultaneously. The progress resulting from selection of superior parents for a given trait depends on heritability, selection differential, and generation interval in years.

Heritabilities of some economically important traits in beef and dairy cattle are presented in Table 4.3. The heritabilities of disease resistance and production traits are presented in Tables 4.4 and 4.5, respectively.

Bulls are evaluated for their ability to transmit the characteristic of high-level milk production both by considering the production level of their ancestors (pedigree) and by considering the production level of their daughters (progeny testing). An index is used as a predictive evaluation of the bull's ability to transmit the characteristic of high-level milk production. Cows and bulls should be selected for structural soundness and functionality.

The important traits for selection of dairy cattle include milk production, milk composition, longevity, and structural soundness. The traits are considered based on their heritability and economic importance. Fertility as a trait has low heritability and therefore can be improved by improving managemental condition. Environment plays significant role on milk production. Therefore, the environmental factors should be standardized to improve milk yield performance through selection. Susceptibilities to cystic ovaries, ketosis, mastitis, and milk fever are having low heritability (5–10%). Traits such as percentages of fat, protein, and solids-not-fat are high in heritability (50%). The traits such as somatic cell counts, udder

Table 4.3 Heritabilities of some economically important traits in beef and dairy cattle

Type of cattle and traits	Heritability (%)
Beef cattle	
Highly heritable traits	
Birth weight	35
Milk production	40
Feedlot gain	40
12 months carcass characteristics	45
Carcass characteristics	40
Age at puberty	40
Scrotal circumference	50
Mature weight	50
Traits of medium heritability	
Weaning weight	25
Carcass yield grade	30
Traits of low heritability	
Calving interval	10
Longevity	0
Dairy cattle	
Highly heritable traits	
Birth weight	50
Butter fat percentage	40
Mature weight	35
Traits of medium heritability	
Milk production	25
Fat production	25
Protein production	25
Excitability	25
Traits of low heritability	
Teat placement	20
Service per conception	5
Mastitis susceptibility	10

Table 4.4 Heritability of disease resistance traits in cattle

Trait	Breed	Heritability	Repetability	*n*	Reference
Susceptibility to nematode parasites (Faecal egg count)	Hereford cattle	0.04 (pre-weaning)		438	[21]
Faecal egg count	Angus cattle	0.32 (post-weaning)		370	[22]
External parasitic infestation (tick count)	*Bos taurus* (Hereford Southline Breed)	0.41			[23]
Mastitis trait	Norwegian red cows	0.09 to 0.11			[24]

Table 4.5 Heritability of production traits of cattle

Trait	Breed	Heritability	Repetability	n	Reference
305-d MY	Ethiopian Holstein	0.15 ± 0.04	0.42 ± 0.02	3733	[25]
LMY	Ethiopian Holstein	0. 17 ± 0.04	0.39 ± 0.02	3552	-do-
LL	Ethiopian Holstein	0.03 ± 0.03	0.12 ± 0.02	2938	-do-
AFC	Ethiopian Holstein	0.47 ± 0.06	–	1125	-do-
CI	Ethiopian Holstein	0.11 ± 0.04	0.22 ± 0.02	2764	-do-
DO days open	Ethiopian Holstein	0.09 ± 0.03 0.	19 ± 0.02	2773	-do-
LL	Frieswal Cattle	0.17 ± 0.10		1249	[26].
CI	Frieswal Cattle	0.11 ± 0.09		1249	-do-
305 MY	Frieswal Cattle	0.51 ± 0.14		1249	-do-
Calving interval (d)	Dairy breeds including Holstein Friesian, Jersey, Montbeliarde, and Norwegian red.	0.03 (0.004)	0.08 (0.004)	142,109	[27]
Days open	-do-	0.05 (0.006)	0.07 (0.007)	97,556	-do-
Calving to first service interval (d)	-do-	0.07 (0.006)	0.11 (0.005)	125,375	-do-
Number of services (U)	-do-	0.02 (0.003)	0.08 (0.003)	127,744	-do-
Pregnancy rate to first service (%)	-do-	0.02 (0.004)	0.04 (0.004)	92,272	-do-

characteristics (depth, milking speed), and reproductive efficiency (56-day non-return rate of daughters) are being included in the selection programme.

The genetic correlation between two traits is indicative of the amount of genetic change in trait A that might be expected from a certain amount of selection pressure applied to trait B. The most important genetic correlations are those that might be associated with milk yields during the first lactation. Fat, solids-not-fat, protein yield, lifetime milk yields, and length of productive life have high genetic correlation with first lactation milk yield (0.70–0.90). Overall type score levelness of rump, udder texture, and strength of fore and rear udder are all negatively correlated with first lactation milk yield (−0.20 to −0.40). Dairy character and udder depth are positively correlated (0.35–0.40) with first lactation yield. Inbreeding tends to increase mortality rate and to reduce all production traits except fat percentage of milk and mature body weight.

4.4.2 Genetic Improvement in Buffaloes

The recording of traits of improved breeding and management strategies for buffaloes was addressed by FAO during 2000. The performance trait recording is being carried out in buffaloes in India, Italy, Bulgaria, Iran, Nepal, and Pakistan. In India, the buffalo genetic improvement programme is promoted by National Dairy Development Board (NDDB) and Indian Council of Agricultural Research, New Delhi. The NDDB supports the development of dairy cooperatives by providing financial assistance and technical expertise and mainly working on Mehsana and Murrah cross buffaloes. Indian Council of Agricultural Research is working under All India coordinated Research project in the home tract of different breeds. In Italy, the buffalo population has increased due to popularity of mozzarella cheese from buffalo milk in European market. The Italian Breeder's association and the buffalo breeder's association provide technical support to carry out selection activities. Similarly, in Egypt, the Ministry of Agricultural and Land Reclamation through the Animal Production Research Institute carries out milk recording and evaluates genetic merit of bulls. In Iran, milk recording and selection are implemented by the Central government through the Animal Breeding Centre of Karaj. Similarly, milk recording activity in buffaloes in Brazil is performed in imported breeds such as Murrah and Mediterranean breed. Bulgaria is also promoting selection of buffaloes by regional agency for selection and reproduction with scientific and technical support from the Agricultural Institute, Department buffalo breeding, Shumen, Nepal is also improving production performance in Murrah breed by the Agricultural Research stations of Nepal Agricultural Research Council. Nepal has also adopted crossbreeding of Indigenous buffaloes with Murrah as a part of genetic improvement Programme. Genetic improvement in improving production performance has been carried out in buffaloes through progeny testing in different countries. The genetic merit of bulls is estimated using BLUP animal model developed from the records of daughters and related animals. The genetic parameters of production traits of buffaloes in different breeds are presented in Table 4.6.

4.4.3 Selection Methods

Methods to establish long-term selection goals require knowledge on genetic and phenotypic parameters of traits that may contribute either directly or indirectly to improve profitability. The parameters are estimated from data derived from parents, progeny, and other close relatives. These include heritabilities, phenotypic and genetic correlations of identified traits. These parameters are estimated from phenotypic and genotypic variances and covariances of important economic traits.

Observed phenotypic variation can be attributed to genetic and environmental attributes. The genetic attributes can be additive or non-additive while the environmental attributes can be permanent or temporary. Dominance variation is observed when two genes (alleles) at a particular locus on a chromosome interact, and one

Table 4.6 Genetic parameters of Production Traits of Buffalo

Trait	Breed	Heritability	Repetability	*n*	Reference
BW	Surti Buffalo	0.188 + 0.112		522	[28]
3 BW	Surti Buffalo	0.175 + 0.108		522	-do-
6 BW	Surti Buffalo	0.216 + 0.122		522	-do-
12 BW	Surti Buffalo	0.144 + 0.096		522	-do-
305-day lactation milk yield	Nili Ravi (Pakistan)	0.01 ± 0.02		3141	[29]
Total lactation milk yield	Nili Ravi	0.001 ± 0.01		3141	-do-
Lactation length	Nili Ravi	0.11 ± 0.06		3197	-do-
Dry period	Nili Ravi	0.07 ± 0.14		1991	-do-
MY	Murrah (Brazil)	0.20 ± 0.01		1578	[30]
AFC	Murrah (Brazil)	0.07 ± 0.05		1218	-do-
IBFSC	Murrah (Brazil)	0.14 ± 0.07		865	-do-
Pregnancy rate (PR)	Murrah (NDRI)	0.02 ± 0.005	0.09 ± 0.04	522	[31]
Test day five Milk yield (TD 5MY)	Murrah (NDRI)	0.12 ± 0.04	0.18 ± 0.05	522	-do-
305 days or less Milk yield (MY)	Murrah (NDRI)	0.17 ± 0.04	0.27 ± 0.04	522	-do-
305 days or less wet average (WA)	Murrah (NDRI)	0.15 ± 0.03	0.26 ± 0.04	522	-do-
BW	Murrah (Ind)	0.12 ± 0.01		590	[32]
3 BW	-do-	0.19 ± 0.02		590	-do-
6BW	-do-	0.22 ± 0.06		590	-do-
9BW	-do-	0.18 ± 0.08		590	-do-
12 BW	-do-	0.20 ± 0.06		590	-do-
Monthly test day **fat** (test day 9 to 296th day)	Murrah buffalo Ind	0.09 ± 0.03		565	[33]
Test day 2 (62th day)	-do-	0.19 ± 0.02		565	-do-
Monthly test day SNF test day 8 (258th day)	-do-	0.06 ± 0.002		565	-do-
Test day 3 (97th day)	-do-	0.21 ± 0.06		565	-do-
Milk yield	Italian river buffalo	0.14		10,663	[34]
Fat yield	-do-	0.11		10,313	
Fat %	-do-	0.17		10,313	
Protein yield	-do-	0.14		9441	

(continued)

Table 4.6 (continued)

Trait	Breed	Heritability	Repetability	*n*	Reference
Protein %	-do-	0.10		9441	
Mozzarella yields	-do-	0.13		9420	

gene completely overrides the effects of another (e.g. the dominance of the polled trait in cattle over horned recessive gene).

The raw material for livestock improvement is genetic variation due to genes that are additive in their effects, and those non-additive gene effects (i.e. epistasis and dominance). Additive gene effect is that which is passed on from one generation to the other. Epistasis is gene action where genes at one locus interact with genes at another locus to cause variation, e.g. the gene that restricts colour in the Charolais cross calves out of Angus dams. The calves are dun in colour because the gene for restriction of colour from the Charolais sire is epistatic to the gene for black colour from the Angus dam.

Physiological effects cause systematic non-genetic variations and adjustment to reduce the non-genetic component of variance. So, measurements of performance may be adjusted for a variety of systematic physiological effects such as age, parity, stage of lactation.

The BLUP animal model method is the most appropriate to rank the animals because of the properties of the predictor and its accuracy. BLUP simultaneously ranks males and females, thus adjusting for non-random mating. This method also identifies all animals by pedigree and can rank the various herds together. This method removes a very high percentage of environmental variance.

Genetic variation decreases as selection progresses and inbreeding may set in if selection goes on for long because the selected population becomes very closely related. The improvement of indigenous breeds through selective breeding, where relevant, should also be given serious consideration. In each country, the policies and practices for delivery of improved genetics and related services to farmers should be formulated in relation to the distribution of cattle population, types of production systems, environmental conditions, availability of resources for livestock production, and the social and economic situation of farmers and people.

The three methods of selection are (1) Tandem, (2) Independent culling level, and (3) Selection index. Economically important traits need to be improved as a single trait at a time or including more than one trait at a time. This depends on correlation between traits and other parameters.

Tandem method: It is the selection—One trait is selected over generation till the desired improvement has been achieved. Subsequently, the second trait is improved. This selection method may be useful for two or three traits that are not correlated. This method can be effective if the situation calls for rapid change in a single, highly heritable trait. For example, in the case of dairy cattle, milk production and milk fat percentage are negatively correlated. If a producer were interested in increasing milk yield, selection pressure solely focused on that trait would eventually result in decreases in percent milk fat.

Independent culling level: In this method, threshold levels need to be defined for each trait under selection programme. The individual, which satisfies the threshold value of each trait, is selected as parents for next generation. Again, this method of selection is useful for three traits only. The method of selection can be implemented without the knowledge of heritability and economic worth. Any bull that does not meet the minimum or maximum level is culled. Birth weight is correlated with calving ease. Birth weights are evaluated on a maximum level (upper limit) since high birth weights results in increased calving difficulty.

Selection Index: It is the most effective selection method. Selection index estimation is possible when genetic parameters and economic worth of the traits are known. The method utilizes the information from individuals, parents, or half sibs or from relatives. This method combines all the information on the basis of which the individuals are selected. This method has some demerit such as it is necessary to maintain all the individuals until the evaluation is completed. The selection index is the most effective system but the most difficult to develop. The disadvantages of this system include the possibility of shifting in the economic value of traits over time and potential failure to identify functional defects.

4.4.4 Crossbreeding

Crossbreeding is used to take advantage of breed complementation and heterosis (hybrid vigour). Breed complementation implies crossing breeds so their strengths and weaknesses complement one another. There is no breed that is superior in all the desired production characteristics; and therefore, planned crossbreeding programmes that use breed complementation can significantly increase herd productivity. The incentive for crossbreeding is the exploitation of hybrid vigour or heterosis, as a result of which the performance of crossbreds exceeds the average of the parental breeds. Heterosis occurs because the parental animals differ in gene composition and that dominant genes carry more favourable effects on traits than do recessive genes.

Heterosis is defined as the increase in productivity in the crossbred progeny above the average of breeds or lines that are crossed. The amount of heterosis expressed is related to the heritability of the trait. The heterosis is highest for low-heritability traits and lowest for high-heritability traits. These relationships are helpful to commercial producers in selecting and crossbreeding to enhance genetic improvement. Selecting genetically superior animals is more important than crossbreeding. There are different systems of crossbreeding such as single cross, rotational crossing, and terminal crossing are applied [35].

Single cross: This is the crossing of any two breeds selected on the basis of their performance traits to produce crossbred offspring with considerable hybrid vigour. Heterosis is fully expressed.

Rotational crossing: A third or fourth breed is systematically introduced into a backcross programme to maintain maximum heterosis. Purebred males are used on crossbred females.

Three-breed terminal cross: The F1 crossbred females are mated to males of a selected third breed and all offspring (F2) slaughtered for meat production. More heterosis can be achieved with this method than with a three-breed rotational cross.

4.4.5 Grading Up

The continuous use of purebred sires of the same breed in a grade herd or flock is called **grading up**. In this situation, grading up is similar to outcrossing. The accumulated percentage of inheritance of the desired purebred is 50% (1/2), 75% (3/4), 84.5% (7/8), and 94% (15/16) for four generations when grading up is practiced. The fourth generation resembles the purebred sires so closely in genetic composition that it approximates the purebred level.

4.4.6 Forming New Lines or Breeds

New breeds have been formed and are currently being formed by crossing several breeds. These are sometimes given a general classification of synthetic breeds and composite breeds. In beef cattle, the Brangus, Barzona, Beefmaster, and Santa Gertrud breeds are composite (synthetic) breeds formed several years ago. MARC I (crosses of Charolais, Brown Swiss, Limousin, Hereford, and Angus breeds) and RX3 (crosses of Red Angus, Hereford, and Red Holstein breeds) are also developed.

4.4.7 Genetic Defects

It is necessary to analyse the hereditary defects in the flock or individual. A single pair of genes that are usually recessive determines most of these defects. When one of these hereditary defects occurs, it is a logical practice to cull both the cow and the bull.

The common genetic defects in cattle today are double muscling, syndactyly (mule foot), arthrogryposis (palate-pastern syndrome), osteoporosis (marble bone disease), hydrocephalus, and dwarfism. Double-muscled cattle usually grow slowly and the fat deposition on the carcass is much less than those of the normal beef animal. **Syndactyly** is a condition in which one or more of the hooves are solid in structure rather than cloven. Mortality rate is high in calves with Syndactyly. **Arthrogryposis** is a defect in which the pastern tendons are contracted, and the upper part of the mouth has not properly fused together. **Osteoporosis** is characterized by the marrow cavity of the long bones being filled with bone tissue. All calves having osteoporosis have short lower jaws, protruding tongue, and impacted molar teeth. A bulging forehead where fluid has accumulated in the brain area is typical of the defect of **hydrocephalus**. Calves with arthrogryposis, hydrocephalus, or osteoporosis usually die shortly after birth.

Some of these abnormalities include achondroplasia (short bones), weavers, limber limbs, rectal-vaginal constriction, dumps, flexed pasterns (feet turned back), fused teats (teats on same side of udder are fused), hairlessness (almost no hair on calf), and syndactylism (only one toe on a foot). Many of these inherited abnormalities are lethal and most are recessive in their mode of inheritance.

4.5 Care and Management

4.5.1 Age and Body Condition Score

The age group of cattle and buffaloes can be determined based on dentition pattern. The dentition pattern is described in Tables 4.7 and 4.8 and body condition score of cattle is described in Table 4.9.

4.5.2 Calf Management

The calves should be managed in better manner to get very good replacement stock. The care during pregnancy period is also very important and should be provided specific health and nutritional management for better foetal growth and immunity. The care and management practices to be followed to avoid embryonic and calf mortality. After parturition, the calf should be wiped with dry cloth to remove mucous from the nose and mouth and artificial respiration should be provided by relaxing the chest with hands. The cow licks the calf immediately after the birth. The mother helps in stimulating breathing and circulation. The naval chord should be tied about 2–5 cm away from the body and cut 1 cm below the ligature with povidone iodine or tincture iodine. The udder should be washed with chlorine solution and dry. The calf should be allowed to suckle the mother and colostrum should be fed to

Table 4.7 Age of cattle by dentition pattern

Type of teeth	Age
No permanent teeth	<2 years
Two permanent teeth	2 years 3 month
Four permanent teeth	3 years
Six permanent teeth	3 years and 6 month
Eight permanent teeth old animals	4 years >4 years

Table 4.8 Age of buffaloes by dentition pattern

Type of teeth	Age
No permanent teeth	2 years
2 large permanent teeth	2 years and 6 month
4 large permanent teeth	3 years and 6 months
6 permanent teeth	4 years and 6 months
8 large permanent teeth Old animals	5–6 years >6 years

Table 4.9 System of body condition scoring (BCS) for beef

Group	BCS	Description
Thin condition	1	*Emaciated*-cow is extremely emaciated, with no palpable fat detectable over spinous processes, transverse processes, hip bones, or ribs. Tail-head and ribs project quite prominently
	2	*Poor*-cow still appears somewhat emaciated but tail-head and ribs are less prominent. Individual spinous processes are still rather sharp to the touch, but there is some tissue cover over dorsal portion of ribs
	3	*Thin*-ribs are still individually identifiable but not quite as sharp to the touch. There is obvious palpable fat along spine and over tail-head, with some tissue cover over dorsal portion of ribs
Borderline condition	4	*Borderline*-individual ribs are no longer visually obvious. The spinous processes can be identified individually on palpation but feel rounded rather than sharp. There is some fat cover over ribs, transverse processes, and hip bones
Optimum moderate condition	5	*Moderate*-cow has generally good overall appearance. On palpation, fat cover over ribs feels spongy, and areas on either side of tail-head now have palpable fat cover
	6	*High moderate*-firm pressure now needs to be applied to feel spinous processes. A high degree of fat is palpable over ribs and around tail-head
	7	*Good*-cow appears fleshy and obviously carries considerable fat. There is very spongy fat cover over ribs and around tail-head. In fact "rounds" or "pones" are beginning to be obvious. There is some fat around vulva and in crotch
Fat condition	8	*Fat*-cow is very fleshy and over-conditioned. Spinous processes almost impossible to palpate. Cow has large fat deposits over ribs and around tail-head, and below vulva. "Rounds" or "pones" are obvious
	9	*Extremely Fat*-cow is obviously extremely wasty and patchy and looks blocky. Tail-head and hips are buried in fatty tissue and "rounds" or "pones" of fat are protruding. Bone structure is no longer visible and barely palpable. Animal's mobility might even be impaired by large fatty deposits

Source: Richards et al. 1986. J. Anim. Sci. 62:300 [36]

new-born. Colostrum feeding is very important as it boosts immunity of new-born and fulfils protein requirement of body. Colostrum contains antitrypsin which avoids digestion of immunoglobulin in the stomach and is absorbed in small intestine. Milk should be fed twice a day to calves and milk feeding should be around 10% of its body weight. The weak calves should be provided milk for 3 times daily. The milk feeding should continue for 6–10 weeks and care should be taken regarding overfeeding of milk to calves. The calves should be given identification number by tattooing in the ear at birth and branding after 1 year. Dehorning may be carried out within 7–10 days after birth with red hot iron or caustic potash stick or electrical method. The fresh water should be provided from 2–3 weeks onwards. The weight of the calves should be recorded at weekly interval up to 6 months of age and monthly interval afterwards to know the growth rate. The housing should be clean, warm, and with proper ventilation. The males should be castrated at the age of 8–10 weeks. Mineral-blocks should be provided for licking to avoid any mineral deficiency. The calves should be provided leafy hay and grain to stimulate rumen development.

4.5.3 Heifer Management

Heifer should be kept in better management to provide high quality replacement stock to the dairy farm. The nutritional requirement should be optimum for normal growth of heifers. During the early stage of growth, the heifer requires relatively more protein than energy; therefore, feed formulation should be prepared accordingly. The growth of heifer depends upon the quality of forage fed. The heifers should be provided better dry shelter with adequate space. The size rather than the age of a dairy heifer at breeding time is important. Small heifers are more likely to have difficulty in calving.

The health care is provided with respect to hygienic housing, water balanced feeding, and preventive care against common diseases. The heifers in the herd should be checked for their proper growth and other progress. Animals lagging behind below the required standards should be removed from the herd. Heifer should be ready for breeding at the age of 15 month with optimum body size. Weight of heifer is an important criterion for deciding breeding as it affects puberty. Heifers with proper body size and weight produce more milk during first lactation.

4.5.4 Lactating Animal Management

The lactating animal should be provided the green leguminous forage ad lib to fulfil the maintenance energy requirement. The concentrate @1 kg for every 2 to 2.5 kg of milk should be provided. Salt and mineral supplements should be given to maintain the lactation.

A cow should come to heat within 3–4 weeks of calving with optimum feeding and care. It is required to maintain individual production records. Water should be provided to drink at frequent intervals. Milking thrice is better than twice since 10–15% more milk can be produced. Cows should be trained to let down milk without calf suckling. This will help to wean the calves early. The animals will get maximum exercise in loose housing system. Grooming of the cows and washing of the buffaloes before milking help in clean milk production. Daily brushing will remove loose hair and dirt from the coat. Grooming will also keep the animal hide pliable. Wallowing of buffaloes or water spraying on their bodies will be detected and care should be taken. It is required to provide at least 60–90 days dry period between calving. If the dry period is not sufficient, the milk yield in subsequent lactation will be reduced. Every animal should be numbered and particulars pertaining to milk, fat%, feed taken, breeding, drying, and calving dates should be recorded.

4.5.5 Dry and Pregnant Animal Management

The usual dry period is about 50-60 days. Short dry periods usually reduce the milk yield in future lactation period as the mammary tissue regeneration and body condition score have not been properly improved. Similarly, long dry period is not advisable as milk yield decreases due to different factors. Dry cows should be separated from pregnant and lactating cows. Dry cows should be fed as per their requirement during different stages of development.

4.5.6 Bull Management

A breeding bull should be housed in "Bull Shed" with sufficient area of floor and proper covering. It is sound practice to provide cool conditions and adequate drinking water. A balanced ration should be fed containing adequate energy, proteins, minerals, and vitamins. Green fodder must be available both before and during breeding season. Bulls should not be kept with lactating cows. Bulls should not be fed diet as provided to lactating animals and the high calcium diet is not appropriate for young bulls.

The breeding bulls should be maintained in good condition for the success of breeding programme. Breeding bull should be put to adequate exercise for large ejaculation containing sperm with higher activity. Adult bulls are required to control properly using nose rings. It is of great importance that males should be fed regularly and not too much at one time, and too little at another. Regular grooming of the breeding bull is practiced. In buffalo bulls regular shaving may be practiced.

4.5.7 Bullock Management

The bullocks should be housed in separate sheds with sufficient space and protection from hot and cool conditions. The animals should have free access to drinking water. Regular grooming of animals should be practiced. Bullocks are normally used for agricultural operations and or transport purpose. The hooves of the bullocks should be provided with metal shoes to protect the hooves from wear and tear. The working hours for bullocks are recommended as follows: Normal Work—6 hours of carting or 4 hours of ploughing. Heavy Work—8 hours of carting or 6 hours of ploughing. The animals may be allowed for grazing and roughages and 1–2 kg of concentrates need to be provided for feeding of bullocks.

4.6 Feeding Requirement

The animal uses feed energy to fulfil basic body function such as maintenance and production. Maintenance energy is used to maintain basal metabolism, to provide for the voluntary activity of the animal, to maintain body temperature and water balance.

Production activities include foetal development, semen production, growth, fat deposition, and production of milk, egg and wool.

Nutrients that contain carbon provide the energy for animals. Carbohydrates, fats, and proteins provide energy. Energy is the force, or power that is used to drive a variety of body systems. Energy needs of animals generally accounts for the largest portion of feed consumed. Several systems have been devised to evaluate feedstuffs for their energy content. Total digestible nutrient (TDN) estimates of feeds were historically the most commonly used energy estimation system. TDN is typically expressed in pounds, kilograms, or percentages after obtaining the proximate analysis and digestibility figures for a feed. The formula for calculating TDN is TDN = (digestible crude protein) + (digestible crude fibre) + (digestible nitrogen-free extract) + (digestible crude fat × 2.25). The factor of 2.25 is used to equate fat to a carbohydrate basis, since fat has 2.25 times energy as an equivalent amount of carbohydrate. There are some shortcomings in using TDN as an energy measurement of feeds, however it works well in balancing rations for cow and buffaloes. However, TDN is being replaced by estimates of **net energy (NE)** in many rations formulation systems. The net energy system (NE) is a more precise energy measurement of feeds. This system usually measures energy values in megacalories per pound or kilogram of feed.

The nutrient requirement varies from growth phase to other physiological stages. There are various nutrient requirements for growth such as energy, protein, minerals, vitamins. The nutrient requirement should be determined for maintenance as well as for milk production and to meet the fat percentage in milk and energy requirement during gestation period. In general the dry matter from roughage should not exceed 2 percent of cow's live weight nor should it be less than 1 percent. The nutrient requirement of cattle is presented in Table 4.10 and the daily nutrient requirements for breeding heifers and cow are presented in Table 4.11.

The ruminants like cattle and Buffaloes meet their protein and energy requirements from fermentation end products (microbial protein and volatile fatty acids). A greater ruminal degradation of both fibre and protein was noticed in buffaloes than in cattle and sheep. This unique ability to better ferment fibre in buffaloes could be the result of adaptation because they have been fed on low-quality high fibrous feeds for longer period. The energy and protein demand of buffaloes are being met by feeding them low-quality roughages, agricultural crop residues, and industrial by-products which contain high levels of lignocellulosic materials, low levels of fermentable carbohydrate and protein.

Dietary rumen degradable protein and rumen undegradable protein does not affect the crude protein digestibility in lactating buffaloes. The concentration of crude protein on dry matter should be between 11 and 14%. Milk yield and milk constituents (fat and protein) yields were greater in buffaloes fed with 50% rumen degradable protein (RDP) than those fed with higher levels of rumen degradable protein (RDP). Buffalo should be provided maintenance requirement and food supplement depending on the dry period and gestation period. Proto [38] considered the nutritional requirements applied to dairy cows adequate also for non-lactating buffaloes and suggested an energy-protein level of 0.65 Milk FU/kg DM and 10.5%

Table 4.10 Daily nutrient requirements of dairy cows

Body weight (lb)	NE_1 (Mcal))	TDN (lb)	Crude protein (lb)	Ca (g)	*P* (g)	Vitamin A (IU)
Mature lactating cows (Maintenance)						
800	7.16	6.9	0.70	0.029	0.024	30
1100	8.46	8.1	0.80	0.044	0.031	38
1300	9.70	9.3	0.89	0.053	0.037	46
1550	10.89	10.5	0.99	0.062	0.044	53
Mature dry cows (last 2 month 3 of gestation)						
800	9.30	9.1	1.96	0.057	0.035	30
1100	11.00	10.8	2.32	0.073	0.044	38
1300	12.61	12.4	2.66	0.086	0.053	46
1550	14.15	13.9	2.98	0.101	0.062	53
Milk production (meal or lb of nutrient per lb of milk for various fat percentages)						
Percentage fat						
3.0	0.291	0.282	0.077	0.0025	0.00170	
3.5	0.313	0.304	0.082	0.0026	0.00175	
4.0	0.336	0.326	0.087	0.0027	0.00180	
4.5	0.354	0.344	0.092	0.0028	0.00185	
5.0	0.377	0.365	0.098	0.0029	0.00190	

Source: National research Council, Nutrient Requirements of Dairy Cattle, 2001 [37]

of crude protein. It was recommended that the amounts of protein in the rations should be higher than 10% during the dry period because with a lower quantity the rumen activity could be compromised [39). During the dry phase, the animals should be fed with fresh forage or hay of good nutritional value and it is advisable to provide 15% DM with a concentrate, therefore re-establishing the reserves of liposoluble vitamins, oligomers and, by means of hydro soluble vitamins, to normalize the rumen fermentation and hepatic functions. Therefore, it is important to feed the dry buffalo as lactating animal and this should be started at least 3 weeks before the presumed parturition. Buffalo utilizes fibres to a large extent than cattle because of large ruminal microbe's population. However, lignin (wood-fibre) is not utilized. The efficiency of fibre digestion is 5–8% higher in buffalo than in cattle. Fat is required in small amounts for the ruminant. Too much unprotected fat in the diet depresses the ability of the microbes to ferment fibres, negatively influencing energy utilization. Minerals and vitamins are required by body for optimum growth and proper muscle and nerve function. They are essential components of body enzymes, hormones, and cells. Minerals play a vital role in improving fertility, production, metabolism, foetus, calf care growth and in increasing immunity against diseases. Vitamins are essential for total body function. Most vitamins are synthesized by the animal or its rumen microbes. Vitamin B is synthesized by ruminal microbes, vitamin K by intestinal microbes, and vitamin C in the tissues. Vitamin D is formed when the precursor, found on the skin on animals and on grass, is exposed to ultraviolet-ray. Vitamins A and E are not synthesized in the animal and should be supplemented. Vitamin A is found in silage, fresh grass, dark green leaves, peas, and

Table 4.11 Daily nutrient requirements (NRC) for breeding heifers and cow

Weight (lb)	Daily gain (lb)	Dry matter consumption (lb)	Total crude protein (lb)	TDN (lb)	ME (Mcal)	Ca (g)	P (g)
Weight heifers-last third of pregnancy							
1,000[a]	0.73	29.7	1.8	11.7	20.9	29	22
1,200[a]	0.88	23.7	2.0	13.3	24.2	33	24
1,400[a]	1.02	26.6	2.2	14.8	26.9	37	27
Cow nursing calves-average milking ability[b] *(first 3 months postpartum)*							
900	0	23.0	1.9	12.5	20.9	24	17
1100	0	26.0	2.1	13.9	23.4	27	19
1400	0	29.0	2.3	15.4	25.7	30	21
Cow nursing calves-superior milking ability[c] *(first 3 months postpartum)*							
900	0	25.4	2.6	14.9	24.9	35	24
1100	0	28.4	2.8	16.4	27.3	37	25
1400	0	31.3	3.0	17.8	29.7	40	27
Dry pregnant mature cows-month 3 of pregnancy							
900	0	16.7	1.2	8.2	13.4	14	14
1100	0	19.5	1.4	9.5	15.6	17	17
1400	0	23.3	1.6	11.4	18.7	21	21
Dry pregnant mature cows-month 8 of pregnancy							
900	0.9	21.0	1.6	10.9	18.3	23	14
1100	0.9	24.1	1.8	12.6	20.9	27	17
1400	0.9	27.0	2.2	14.2	23.9	32	21

Source: National Research Council, *Nutrient Requirements of Beef Cattle,* 2000 [37]
[a]Mature weight potential
[b]Ten pounds of milk per day (equivalent of approximately 450 lb. of calf at weaning if there is adequate forage)
[c]Twenty pounds of milk per day (equivalent of approximately 650 lb. of calf at weaning if there is adequate forage)

carrots. Cereals are a source of vitamin E. Energy and protein requirements for the production of 1 kg of buffalo milk relative to the fat and protein content are presented in Table 4.12. Indicative characteristic of requirements of the dry buffalo herd is presented in Table 4.13.

4.7 Housing

Housing is required to protect livestock from extreme heat, direct sunshine, cold, and rain. The ruminants produce efficiently if proper shelter is provided. The housing requirement will be different for different ecological conditions. Therefore, the factors such as temperature, humidity, rainfall pattern, wind flow direction should be considered before designing house. Loose housing with free stalls (cubicles) systems are entirely satisfactory in semiarid areas. Open type of sheds has advantage over closed type shed. Mean temperature and minimum temperature in close shed

Table 4.12 Energy and protein requirements for the production of 1 kg of buffalo milk relative to the fat and protein content [38]

Energy requirements (Milk Fu/kg of milk)														
Milk fat		6.5	7.0	7.5	8.0	8.3	8.5	9.0	9.5	10.0	10.5	11.0	11.5	12.0
NEL		0.61	0.64	0.67	0.70	0.720	0.73	0.76	0.79	0.82	0.85	0.87	0.90	0.93
Protein requirements (g/ kg of milk)														
Milk protein (%)	3.5	3.7	3.9	4.1	4.3		4.5	4.7	4.9		5.1	5.3	5.5	
CP	99	105	111	116	122		128	134	139		145	151	157	

Table 4.13 Indicative characteristic of requirements of the dry buffalo herd [39]

Dry matter (kg)	10.12
NE_L (Milk FU/kg DM)	0.63–0.65
CP (% DM)	10–11
NDF (% DM)	52–58
Starch + sugar (% DM)	8–10

were significantly higher than those of open type shed. The close type shed significantly contributed to higher ambient temperature during both hot dry and hot humid period. The mean vapour pressure in close type shed has been reported to be higher than open type shed.

The orientation of shade is also an important aspect of shed design. The orientation of livestock shed with its long axis running east-west provides a cooler environment than one with a north-south orientation [40]. In the East-West oriented shelters, animals get more opportunity for radiation exchange with cooler north sky. The shelter is shaded for a greater part of the day resulting into lower floor temperature. With the east-west orientation, the feed and water troughs can be under the shade, which will allow the animals to eat and drink in shade at any time of the day. However, the shaded area should be increased to 3–4 metres per cow. With the north-south orientation, the sun will strike every part of the floor area under and on either side of the roof during the day. This will help to keep the paved area dry. A shaded area of 2.5–3 metres per cow is adequate if feed and water troughs are placed away from the shaded area. If yard space is limited and only 4–5 square metres per cow is available, then concrete paving is highly desirable.

Roofing material may be hay or straw, galvanized steel, plywood, and several types of plastics are the roofing material. A 4–6-inch-thick hay thatch does not receive much heat from the upper surface by conduction. Hay thatch is more suitable for hot dry climate than hot humid climate (need for frequent removal of thatch is limiting factor). Wood makes good shed material but cracks develop easily and needs treatments frequently. Surfaces around shelter are very important in view of radiation exchange between different surfaces and the shelter. The temperature of different surfaces varies significantly at same air temperature. It is clear that green surface does not heat up as much as other surfaces like gravel or loose loam. It should be white outside and coloured inside Reflectivity of white colour is around 75%. The reflectivity of the underneath surface should be less as it determines the quantity of incidental energy from the ground which will be reflected back down to the animals.

Floor space requirement for calf is 1.5 to 2 m^2, Adult male 7.0 m^2 and for adult female is 4.0–5.0 m^2. The optimum width of the shelter is 5.0–6.0 m. The cow stand should be of minimum 5.5 × 9.0 feet with Pucca and well drained floor and shaded resting area of about 30–40 square feet/animal. The minimum roof height should be 10.0 feet to reduce the heat load. The height of shelter in hot climate should be between 3.0–5.0 m. A height less than 3.0 m interferes with proper ventilation resulting into reduced convective heat loss from animals.

A bull pen should have a shaded resting area of 12–15 m^2 and a large exercise area of 20–30 m^2. The walls of the pen must be strong. The gate must be properly designed and must have at least two exits where the herd worker can move. The stall can have ramps at the sides to support the bull's front feet. The suitable housing for calves should be provided to protect from climatic stress, infections, and parasites. Individual pens for calves from birth to 3 months of age are often built with an elevated slatted floor. The floor should be always dry and clean.

The housing condition for medium to large herds should be designed to cater the need of animal and farm operation. A medium- to large-scale dairy unit may include the following facilities: 1. Resting area for cows: (a) paved shade; or (b) deep bedding in an open-sided barn; or (c) free stalls in an open-sided barn. 2. Exercise yard (paved or unpaved). 3. Paved feed area: (a) fence-line feed trough (shaded or unshaded); or (b) self-feeding from a silage clamp. 4. Milking Centre: (a) milking shed or parlour; and (b) collecting yard (part of the exercise yard); and (c) dairy, including milk store; and (d) motor room. 5. Bull pen with a service stall. 6. Calving pen(s). 7. Calf accommodation. 8. Young stock accommodation (yard with paved shade and feed area). 9. Bulk feed store (hay and silage). 10. Concentrate feed store. 11. Veterinary facilities: (a) diversion pen with artificial insemination stalls; and (b) isolation pen. 12. Waste stores: (a) slurry storage; or (b) separate storage of solids and effluents. 13. Office and staff facilities.

Each of the parts of the dairy unit may be planned in many different ways to suit the production management system and the chosen method of feeding.

4.8 Health Care at Different Stages of Growth and Development

4.8.1 Maintaining Biosecurity

Biosecurity measures are part of management system to prevent the spread of infectious diseases by minimizing the movement of biological organisms within and between groups of animals. Biosecurity is becoming part of effective livestock management and also required to meet the standard of world trade in livestock. The development of biosecurity plan becomes critical where animals are intensively managed.

The components of an effective biosecurity system include herd health management, serological testing, treatment and vaccination protocols, monitoring incidence of disease, and record keeping. Biosecurity plan also specifies design of isolation facilities, duration of isolation, protocols for employees working in the isolation units to prevent contamination of other farm sites, and additional elements of a critical control point management system. The biosecurity plan will also detail assuring integrity of farm perimeters, managing farm employees and visitors, pest management, quality assurance of feed and water supplies, managing outside trucks used to transport animals to and from the farm, disposal of dead and chronically sick animals, and manure disposal. It is highly essential not to introduce of diseased or infected animals into your flock. The animals should be purchased with known history and parent details. The animals should be screened for certain transmissible diseases and kept for quarantine for 30 days or more. The biosecurity measures should define isolation protocol, disposal of dead animals, and disposal of waste materials from farm. The animals should be protected from rodents, birds, and other agents. The outside vehicles should not be allowed to livestock areas and protective clothing should be provided to visitors.

4.8.2 Diseases

The major three basic approaches are prevention, control, and eradication in dealing with any disease. The major diseases of cattle and buffaloes are described.

4.8.2.1 Anthrax

It is an acute infectious disease with zoonotic importance. The disease affects mainly cattle, buffaloes, and sheep. The disease is caused by *Bacillus anthracis,* and outbreaks are seen in warm and humid conditions. It is transmitted through ingestion of contaminated feed, inhalation, or via skin cuts. Cattle, buffaloes, and sheep suffer from per acute or acute form of the disease. In per acute form, animals die suddenly within 2 hours with release of blood from natural openings like mouth, nostrils, and anus. Fever, muscular tremors, dyspnoea, and convulsions are recorded prior to death. In acute disease, animals survive for 3–4 days and reveal hyperpyrexia, depression, and oedematous swelling of tongue and throat. The mucous membranes are congested and reveal haemorrhagic spots. The precaution is not to open the dead carcass for the post-mortem. The disinfection of shed should be done by using 5–10% caustic soda or 5% phenyl or 2–4% formalin. Animal movement should be restricted during disease outbreak.

4.8.2.2 Haemorrhagic Septicaemia

The disease is caused by bacteria *Pasteurella multocida* and outbreak occurs during rainy season. Cows, buffaloes, sheep, goat, and camel are mostly affected. It is an acute septicaemic disease characterized by high fever, bronchopneumonia, respiratory distress, swelling on throat and brisket, salivation, and petechial haemorrhage on sub mucosae. The contaminated feed should be disposed properly. The vaccination should be carried out in flock before start of rainy season.

4.8.2.3 Tuberculosis

It is a chronic, wasting, infectious disease of cattle, sheep, goat, pig, and other mammalian species and birds. The disease in cattle is mainly caused by *Mycobacterium bovis*. It is a zoonotic disease and can be transmitted to human being via milk. The progressive weakness is observed in animals. Isolation and culling of infected animals should be carried out. It is considered a disease of socio-economic and public health importance as *M. bovis* is a zoonotic bacterium and it is the major cause of human infection in developing countries [41].

Tuberculosis lesions can be classified as acute miliary, nodular lesions, and chronic organ tuberculosis. In the pulmonary form a low-grade fever, loss of appetite, emaciation, chronic intermittent hacking cough and associated pneumonia, difficult breathing, weakness, dryness of skin are observed. The swelling of supramammary lymph nodes are observable when mammary glands are affected. The incidence of human tuberculosis caused by *Mycobacterium bovis* has markedly dropped with the pasteurization of milk. It has also dropped in the areas where tuberculosis eradication programmes have been carried out.

4.8.2.4 Black Quarter

It is an acute infectious disease characterized by severe toxaemia and gangrenous swelling of muscles. The disease is caused by bacteria *Clostridium chauvoei*. It is a soil-borne infection and spreads mainly through ingestion or skin abrasions during docking and shearing. The disease is mostly occurring in cows, buffaloes, sheep, and goat of 6 months to 2 years of age. In per acute case animal is found dead without showing any symptoms within 24 hrs. Swelling occurs in muscles of shoulder, neck, hip, chest, thigh, and quarter. The affected area becomes dark and gases are released by puncturing swollen area. The vaccination should be carried out before rainy season.

4.8.2.5 Mastitis

It is the inflammation of udder and occurs in cattle, buffaloes, sheep, and goat. The major causes of the infection are unhygienic environment, incomplete milking, and injury to teat and udder. In neglected cases, udder becomes fibrosed and refractory to treatment. Systemic and local antibiotics are administered along with anti-inflammatory drugs. It is necessary to maintain strict hygienic conditions in the flock.

4.8.2.6 Brucellosis

It is a contagious zoonotic disease primarily of cattle, sheep, goats, and pigs. The disease is characterized by abortion in late pregnancy. The disease is caused by *Brucella abortus*. In bulls, testicles are enlarged and necrosis occurs. Brucellosis in animals is primarily a reproductive disease characterized by abortion, retained placenta, and impaired fertility in the principal animal host. It has been established that brucellosis in bulls does not always result in infertility, although semen quality may be affected. Bulls that remain fertile and functionally active will shed Brucella organisms with the semen during the acute phase of the disease. Mild cases are characterized by synovitis and painful swelling of affected joints.

The disease is also known as Bang's disease, Malta fever, and undulant fever (in man). Buffaloes are very susceptible to *Brucella abortus*. Abortions in cattle caused by *B. abortus* and usually occur before the fourth or fifth month of pregnancy. Pregnant females are more likely to become infected than non-pregnant cattle or males. The oral route transmission of *B. abortus* occurs as cattle tend to lick aborted foetuses and the genital discharge of an aborting cow. The infection usually occurs through ingestion of feed and water contaminated with the uterine discharges of an aborted animal or its foetus. The organism is also spread via semen from infected bulls or contaminated udder during milking.

4.8.2.7 Foot Rot

It is an infectious disease of ruminants characterized by inflammation, necrosis, and ulcerations in interdigital space and coronary band. The disease is caused by *Fusobacterium necrophorum.* The affected area should be washed with antiseptic solution and antibiotic cream may be applied topically.

4.8.2.8 Calf Scours

Calf scours is the disease of new-born calves below 3 months age. It is mainly caused by *E. coli.* The disease is characterized by profuse yellowish diarrhoea, rise in body temperature, anorexia, dehydration, sunken eyes, and rough hair coat. Calves become progressively more dull, listless, weak, and prostrate. In kids/lambs, the symptoms noticed below 1 month of age group. It is required to adopt hygienic measures in the shed and handling of calves.

4.8.2.9 Listeriosis

It is an infectious disease of bovines as well as sheep and goats. The disease is caused by *Listeria monocytogenes.* The disease is also known as circling disease. There is moderate to high fever, loss of appetite, and animals press head to a fixed object. In nervous form, star gazing posture is observed in sheep and goats. The disease mainly spreads via wild animals and rodents. It is necessary to adopt strict hygiene conditions at the farm to reduce the occurrence of disease.

4.8.2.10 Actinomycosis

It is a chronic infection of cattle and characterized by rarefying osteomyelitis of skull bone. The disease is caused by *Actinomyces bovis*. Mostly the bovines of 2–5 years of age are affected. A circumscribed protuberance at mandible or maxilla is observed and the rarefaction and destruction of bones (Lumpy Jaw) observed. The iodine and sulphur have been found useful to treat the disease.

4.8.2.11 Actinobacillosis

It is a chronic disease occurring sporadically and characterized by inflammation of tongue and soft tissues. The disease is caused by *Actinobacillus lignieresii*. The disease occurs in cattle, buffaloes, and sheep. The younger animals of 2–5 years of age are more susceptible. In bovines, tongue becomes enlarged, hard, and protruded outside. Intermandibular space is swollen and hard and mandibular lymph nodes are enlarged. Tongue is usually not involved to that extent in sheep and lesions are seen in lower jaw, face, nasal passage. Potassium iodide orally or sodium iodide intravenously has also been found effective especially if used at initial stages.

4.8.2.12 Paratuberculosis

It is an important chronic infectious disease of ruminants and characterized by enteritis and progressive weakness. The disease is caused by *Mycobacterium paratuberculosis*. The disease is observed in 2–6 years old animals as the organisms grow very slowly. In sheep and goats' faeces is soft but diarrhoea is usually not seen. It is difficult to control the disease since it has long incubation period. Eradication of infected animals and carriers and maintenance of strict hygienic conditions are recommended.

4.8.2.13 Foot and Mouth disease (FMD)

Foot and Mouth Disease (FMD) is caused by the Aphthovirus, an RNA virus pertaining to the Picornaviridae family that affects cloven hoofed animals. It is an acute highly contagious viral disease resulting in formation of vesicles in the mouth and feet. It is mostly affected in cattle, buffalo, sheep, and goat. It is usually occurring in rainy season (July to October) and winter (January to February), but may occur at any month of the year. Crossbreds and young animals are severely affected and more susceptible. The main symptoms are high fever, depression and dullness, appearance of blisters or vesicles in the tongue, lips, gums, dental pad, palate, muzzle, teats, and on feet. The clinical signs in sheep and goats are less evident.

Asia 1 is more severe in buffaloes than in cattle [42]. Foot and mouth disease has been described in Indian buffalo (*Bubalus bubalis*) with the same features as that in cattle in terms of temperature, viraemia, virus replication in pharyngeal area, excretion, and persistent infection. The presence of minor tongue lesions and initial scaly foot lesions eventually becoming vesicular in buffalo [43]. The tongue and feet should be washed with mild disinfectant (potassium permanganate 0.001%, caustic soda 2%, soda ash 4% or sodium-bi-carbonate 0.1–5%, or 1% alum solution).

4.8.2.14 Malignant Catarrhal Fever

Malignant catarrhal fever (MCF) is a generalized viral disease caused by lymphotropic herpes virus. Domestic and wild sheep and goats are also considered reservoir hosts for the MCF virus. Sheep-associated MCF occurs worldwide. The clinical signs in domestic cattle and buffaloes and in many species of wild ruminants are characterized by high fever, profuse nasal discharge, corneal opacity, ophthalmia, generalized lymphadenopathy, leukopenia, and severe inflammation of the conjunctival, oral, and nasal mucosa with necrosis in the oral and nasal cavities sometimes extending into the oesophagus and trachea.

4.8.2.15 Bovine Viral Diarrhoea

The BVD virus belongs to the Pestivirus genus (Flaviviridae family). The bovine viral diarrhoea (BVDV) infection is a major worldwide problem affecting different species of ruminants [44]. The significant economic impact inflicted is due to productive and reproductive losses: by reduced milk yield, reduced conception rate, abortion, foetus mummification, congenital malformations, weak calves, and increased animal mortality. The other species that have been infected with BVDV include sheep, goats, lamas, pigs, giraffe, captive and free-ranging deer, antelope, elk, buffalo, water buffalo, reindeer, and wildebeest.

BVDV infection can be described as an acute, chronic, persistent, mucosal disease and haemorrhagic syndrome. In buffaloes, clinical signs are mild compared to those in cattle, but they are similar in chronology.

4.8.2.16 Rabies

It is an acute viral disease and caused by rhabdovirus and affects the nervous tissue. There is an excessive salivation, hydrophobia and paralysis and recumbency. The infection spreads mostly by saliva through contamination of wounds or bites of infected animals. The buffaloes defend themselves well from rabid animals and no cases of rabies have been found. The disease may present a paralytic (drooling of saliva, eructation, grinding of teeth, tail movement, anorexia, stiffness of hind limbs, paralysis and recumbency, death in 2–3 days) or furious form (alert state, hypersensitivity, sexual excitement, inability to swallow, ramming of head on fixed objects, loud bellowing, collapse, and death) [45]. Wounds should be irrigated with a soap solution and water. Post exposure vaccination can be performed. Suspected animals should be kept under close observation avoiding euthanasia. Vaccination of cats and dogs are important.

4.8.2.17 Buffalo Pox

It is caused by orthopox virus and the virus affects buffaloes only. It has been reported in India, Indonesia, Italy, Pakistan, Russia, and Egypt [46]. The disease can be observed in both localized and generalized form. The pox skin lesions are mainly on the teats, udder, and thighs. Signs of the disease are represented by fever, anorexia, dullness, depression, and congestion of conjunctivae [45].

4.8.2.18 Babesiosis

It is tick-borne protozoan disease of exotic and crossbred cattle caused by *Babesia* parasites. It is transmitted by tick's bite. The disease is characterized by high fever and haemoglobinuria. The animal produce coffee coloured urine. It is required to control ticks by spraying of insecticides on animal body and cattle shed at every 3-month interval.

4.8.2.19 Theileriosis

It is a tick-borne disease of exotic and crossbred cattle caused *Theileria parva* and *T. annulata*. It occurs generally during summer and rainy season. The young calves are susceptible as compared to other age groups. The disease is characterized by high fever, loss of appetite, and difficulty in breathing. It is required to control ticks by spraying of insecticides on animal body and cattle shed for every 3-month interval.

4.8.2.20 Trypanosomiasis

It is an infectious disease of bovine due to several Trypanosome sp. The disease is characterized by intermittent fever, progressive anaemia, and emaciation.

4.8.2.21 Coccidiosis

The disease is caused by a protozoan that affects the calves that are 6–12 months of age and causes significant economic loss. The disease is caused by *Eimeria zuernii* and *Eimeria bovis*. The major symptoms include diarrhoea, rough coat, loss of appetite, and weight loss. Calves acquire infection as soon as they begin grazing or eating food other than their mother's milk.

4.8.2.22 Mad Cow Disease (BSE)

Bovine Spongiform Encephalopathy is also known as BSE. The disease is most commonly referred as Mad Cow Disease. BSE is caused by prions. BSE is a progressive degenerative disease that affects the central nervous system of cattle. BSE has an incubation period of 4–5 years. BSE is fatal for cattle and death results within weeks to months of its onset. The disease is characterized by nervousness or aggression, change in attitude and behaviour, abnormal posture, coordination problems, difficulty in walking or getting up off the ground, reduced milk production, acute muscular twitching. BSE-affected animals resemble those of rabies.

4.8.2.23 Leptospirosis

Leptospirosis is a zoonotic disease in almost all parts of the world. The disease is caused by the genus Leptospira. It is an acute, often fatal, disease characterized by haemorrhage, haemoglobinuria, and icterus. Non-fatal infections are often characterized by fever, anaemia, abortions, sterility, decreased lactation, and mastitis.

4.8.3 Metabolic Disorders

4.8.3.1 Milk Fever

It is metabolic disorder of adult females and occurs in high milk producing animals. It is more common during first week of lactation but may occur any time of lactation. The animal exhibits weakness, not able to stand, muscle tremors, absence of pupillary light reflex, kinking of the neck, and lateral recumbency. It is necessary to provide low calcium and high phosphorus diet during last 1–2 weeks of gestation, provide calcium rich mineral mixture after calving.

4.8.3.2 Bloat

It is abnormal distension of the rumen and reticulum. The disease is characterized by enlarged abdomen (left flank), difficulty in breathing, abdominal pain, rolling on ground, frequent urination and defaecation, tympanic sound on percussion. The condition happens due to sudden change in feed, feeding of spoiled feed and fodder, feeding of kitchen waste in large quantity. The animal should be treated immediately and in severe case left flank should be punctured with warm pointed object. It is required to avoid sudden change in feed.

4.8.3.3 Ketosis

Ketosis or acetonaemia is a common metabolic disease of lactating cows. Ketosis is mainly observed during the first 10 to 60 days after calving in high-yielding cows. The disease is characterized by lowered blood sugar in the circulating blood, which causes the formation and release of ketone bodies. Ketone bodies (specifically acetone) are responsible for the "sweetish" smell detectable on the breath, and in the milk or urine of affected cows. Symptoms of ketosis in dairy cattle include dullness, depression, a staring expression, rapid loss of weight, a drop-in milk production, constipation, mucus covered faeces, incoordination, and partial paralysis. The prevention of ketosis includes the addition of sodium propionate and propylene glycol to the dairy ration.

4.8.3.4 Hypomagnesaemia

The magnesium ion is essential for normal bone metabolism, normal nerve function, and muscle irritability. Magnesium also plays an essential part in the coenzyme system which links normal carbohydrate metabolism with phosphate metabolism and the provision of energy for muscle contraction. A deficiency of magnesium occurs commonly in cattle.

References

1. Sethi RK (2003) Buffalo breeds of India. In: Proc of Fourth Asian Buffalo congress, New Delhi, India, 25 to 28 Feb
2. Mudgal VD (1988) Proceedings of the second world Buffalo congress, New Delhi, India, held during 12-17th December, 1988, p 454
3. Di Berardino D, Iannuzzi L (1981) Chromosome banding homologies in swamp and Murrah buffalo. J Heredity 72:183–188
4. Alexiev A (1998) The water Buffalo. St. Kliment Ohridski University Press, Sofia
5. Tunio AN (1999) Personal communication
6. Latifova E (2000) Case study on buffalo recording. Breeding and management strategies of buffalo in Azerbaijan. In: Proc. workshop on animal recording and management strategies for buffaloes. ICAR technical series 4, pp 45–50
7. Turabov T (1991) Selection of Caucasian breed of buffalo. PhD thesis, Academy of Agriculture, Baku, Azerbaijan
8. Turabov T (1997a) The modern condition of buffalo production in Azerbaijan and the ways for further improvement. In: Proc. of the fifth buffalo congress, Caserta, pp 884–888
9. Turabov T (1997b) Buffalo population and production in Azerbaijan. Buffalo Newsletter 8:9–10
10. Naderfard HR, Qanami A (1997) Buffalo breeding in Iran. In: Proc. fifth buffalo congress, Caserta, pp 942–943
11. Marmarian Y (2000) Buffalo breeding in Armenia. In: Proc. workshop on animal recording and management strategies for Buffalo. ICAR technical series, vol 4. pp 73–76
12. Qanemi A (1998) Buffalo population and production in Iran. Buffalo Newsletter 10:12–14
13. Borghese A (2005) Personal information
14. Peeva T, Vankov K, Vasilev M (1991) Persistency index of lactation milk production in Buffalo cows. In: Proc. of the third world buffalo congress, agricultural academy, Varna, Sofia, p 61
15. Peeva T (1996) Buffalo population and production in Bulgaria. Buffalo Newsletter 4:6–7
16. El-Kirabi F (1995) Buffalo population and production in Egypt. Buffalo Newsletter 3:8–9

17. Nigm AA (1996) Characterization of Egyptian Buffalo. In: Proceedings of international symposium on buffalo resources and production systems. Giza, Egypt, 17 Oct
18. Ragab MT, Abdel-Salam MF (1963) Heredity and repeatability of bodyweight and growth rate in Egyptian buffalo and cattle. Egypt J Anim Prod 3:15–26
19. Mokhtar SA (1971) Study on some economic traits of Holstein Friesian cattle in U.A.R. MSc thesis, Faculty of Agriculture, Ain Shams University, Cairo, Egypt
20. Hammond K, GASSER HU, Mcdonald CA (1992) Animal breeding, the modern approach, post graduate foundation in veterinary science. University of Sydney, Sydney
21. Barlow R, Piper LR (1985) Genetic analyses of nematode egg counts in Hereford and crossbred Hereford cattle in the subtropics of New South Wales. Livest Prod Sci 12:79–84
22. Morris CA, Green RS, Cullen NG et al (2003) Genetic and phenotypic relationships among faecal egg count, anti-nematode antibody level and live weight in Angus cattle. Anim Sci 76:167–174
23. Henshall JM (2004) A genetic analysis of parasite resistance traits in a tropically adapted line of Bos taurus. Aust J Agric Res 55:1109–1116
24. Heringstad B, Chang YM, Gianola D et al (2005) Genetic analysis of clinical mastitis, Milk fever, ketosis, and retained placenta in three lactations of Norwegian red cows. J Dairy Sci 88:3273–3281
25. Wondossen A, Mohammed A, Negussie E (2017) Estimation of genetic parameters of the productive and reproductive traits in Ethiopian Holstein using multi-trait models. Asian Aust J Anim Sci 30(11):1550–1556
26. Kakati P, Panchal D, Patel AC et al (2017) Genetic parameters of production and reproduction traits and factors affecting it in Frieswal cattle. Int J Livestock Res 7(7):190–199
27. Carthy TR, Ryan DP, Fitzgerald AM et al (2015) Genetic parameters of ovarian and uterine reproductive traits in dairy cows. J Dairy Sci 98:4095–4106
28. Pandya GM, Joshi CG, Rank DN et al (2015) Buffalo Bull 34(2):189–195
29. Thevamanoharan K, Vandepitte W, Mohiuddin G et al (2002) Animal model heritability estimates for various production and reproduction traits of Nili-Ravi buffaloes. Int J Agric Biol 4(3):357–361
30. Seno LO, Cardoso VL, El Faro L et al (2010) Genetic parameters for milk yield, age at first calving and interval between first and second calving in milk Murrah buffaloes. Livestock Res Rural Dev 22(2)
31. Jamuna V, Chakravarty AK, Singh A et al (2015) Genetic parameters of fertility and production traits in Murrah buffaloes. Indian J Anim Res 49(3):288–229
32. Thiruvenkadan AK, Panneerselvam S, Rajendran R (2009) Non-genetic and genetic factors influencing growth performance in Murrah buffalos. S A J Anim Sci 39(Suppl 1)
33. Jamuna V, Gupta AK, Chakravarty AK et al (2017) Estimation of genetic parameters for milk composition traits in Indian Murrah buffaloes. Adv Anim Veterinary Sci 5(5):229–233
34. Rosati A, Dale Van Vleck L (2002) Estimation of genetic parameters for milk, fat, protein and mozzarella cheese production for the Italian river buffalo Bubalus bubalis population. Livestock Prod Sci 74(2):185–190
35. Maree C, Casey NH (1993) Livestock production systems, principles and practice. Agri Development Foundation, Brooklyn
36. Richards MW, Spitzer JC, Warner MB (1986) Effect of varying levels of postpartum nutrition and body condition at calving on subsequent reproductive performance in beef cattle. J Anim Sci 62:300–306
37. National Research Council (n.d.) Nutrient Requirements of Beef cattle (2000); Dairy Cattle (2001); Goats (1981); Horses (1989); Poultry (1994): Sheep (1985); and swine (1998). National Academy Press, Washington
38. Proto V (1993) L'alimentazionedellabufala. Giornata di studio "Alimentazionezootecnicaequalità del latte bovino e bufalino". Eboli (SA). pp 1-42
39. Bertoni G, Te DL, Bartocci S (1994) Nuoveacquisizioninel campo dell'alimentazionedeibufali. Agric Ricerca 153:159–172

40. Kelly CF, Bond TE, Ittner NR (1950) Thermal design of livestock shelters. Agric Eng 31:601–606
41. Tariq Javed M, Usman M, Cagiola M, et al (2004) A study on tuberculosis in buffaloes: some epidemiological aspects along with haematological and serum protein changes. In: Proceedings of the seventh world buffalo congress, 20–23 October, Manila, pp 73–83
42. Hajela SK, Sharma SK (1978) Indian Veterinary Med J 2:103–104
43. Gomes I, Ramalho AK, De Mello PA (1997) Infectivity assays of foot-and-mouth disease virus: contact transmission between cattle and buffalo (Bubalus bubalis) in the early stages of infection. Vet Rec 140(2):43–47
44. Pringle CR (1999) The universal system of virus taxonomy, updated to include the new proposal ratified by the international committee on taxonomy of viruses during 1998. Arch Virol 144:421–429
45. Adlakha SC, Sharma SN (1992) In: Tulloh NM, Holmes JHG (eds) Infectious diseases: worlds animal science C6. Buffalo production. Elsevier, Amsterdam, pp 292–298
46. Sharma MC, Mahesh Kumar (2003) Infectious diseases of buffaloes. In: Proceedings of the fourth Asian Buffalo congress, 25–28 Feb, New Delhi, India, pp 152–168

Factors Influencing Livestock Way of Life

5

5.1 Climate Change on Livestock

Livestock production is a key component of world agriculture. Livestock production system directly supports the livelihoods of 600 million smallholder farmers in developing countries [1]. The human civilization is dependent on animal agriculture to fulfill the requirement such as protein, fibre, leather, transport, draft, and manure. The demand for livestock product is increasing in both developed and developing countries and therefore industrial livestock production is being adopted in different countries to fulfill the requirement. Therefore, it is necessary to maintain livestock productivity in different production system and to analyse the effect of climate, disease, and nutrition on productivity. Moreover, it is essential to maintain genetic diversity, conserve the biodiversity, and bring sustainability to livestock production with respect to environmental condition and geography.

Animal productivity has been optimized in a narrow environmental condition. Animal production efficiency is compromised by increase or decrease of temperature and affects several productive parameters such as body growth, milk yield, reproduction, and carcass traits. Heat or cold stress is one of the factors which limit the productivity of animal in different climatic conditions, and the problem is more evident in high productive animals. Basal or metabolic heat production increases with enhanced milk production [2]. Therefore, traditional selection for increasing productivity also makes animal more prone to environmental stress and disease susceptibility. It is necessary to understand how the climate variability affects the animal performance and devising future strategies to maintain the animal well-being, performance, and economics.

Climate change is the major concerns that will affect livestock productivity globally and will have significant effect on specific genotype and disease trend over the region [3]. The tropical region will be severely affected by climate change and require the genotypes which have better adaptability. Therefore, the tropical region requires small size heat tolerant animals which can survive longer in hot weather with limited feed and water. Similarly, there will be increase in disease and

P. K. Rout, B. K. Behera, *Sustainability in Ruminant Livestock*,
https://doi.org/10.1007/978-981-33-4343-6_5

parasite problem in temperate and Mediterranean climate. Therefore, the livestock production in temperate climate will shift towards genotypes having higher tolerance to disease and parasites.

5.1.1 Adaptation and Livestock Productivity

Adaptability of an individual can be defined as the ability to survive and reproduce within a defined environment or degree to which an organism, population, or species can remain/become adapted to a wide range of environments by physiological or by genetic means [4]. Livestock production is going to face heat stress, water scarcity, and coastal flooding due to climate change. Therefore, it is required to evaluate the different genotypes in varied locations for national and global application in future. Ruminants have evolved and adapted to all diverse environments and carry unique combinations of genes that define productive and adaptive capabilities. The ruminants have wider genetic base as it is distributed over varied geographical and climatic conditions with differential production capability. Due to wider genetic variability, the ruminants can be manipulated to develop suitable adaptable animal in relation to different ecotypes.

Climate stress affects animal bioenergetics, and has a negative impact on animal performance and well-being. Nutrition also affects livestock productivity during varied climatic condition. In tropical climate, dry season is characterized by scanty rainfall. In Mediterranean climate, there is a very scanty rainfall during summer season. During the dry period, the pasture availability is less and the pastures have high fibre and low-protein content. Therefore, the poor nutrition of livestock increases the susceptibility to diseases and the animal becomes poor converter of biomass and hampers the productivity.

The world climate has been classified under the following six categories such as tropical (group A), dry (group B), temperate (group C), continental (group D), polar (group E), and alpine (group H). Tropical climate is usually divided into three sub-groups such as tropical rain forest, Monsoon and Tropical wet and dry, or savannah. The tropical climate is characterized by dry and rainy season with different durations according to geographical locations.

In tropical climate, the livestock species have developed the several physiological strategies to cope with environmental conditions. The physiological strategies such as higher surface skin area localized fat depots and some behavioural adaptation. Domestic cattle comprise two species mainly *Bos taurus* (European cattle) and *Bos indicus* (Indian cattle). *Bos indicus* cattle are adapted to tropical climates, and *Bos taurus* cattle are adapted to temperate environment. The heat stress adaptation of zebu cattle has been described [5], and the differential adaptation to high ambient temperature by zebu and *Bos taurus* cattle has been analysed at field level [6, 7], environmental chamber [8], and gene expression studies [6, 7].

It is required that the amount of heat produced by the body must equal to the amount dissipated to the surrounding environment. *B. indicus* is more adaptive to harsh environment due to the following attributes such as increasing surface area per unit of body weight; increasing temperature gradient between animal and air; increasing conduction of heat from the body core to the skin; decreasing solar radiation reflection; increasing metabolic rate and feed intake; and adjusting cellular mechanisms. *Bos indicus* due to their physiological abilities to maintain body temperature and modifying cellular stress regulation remains productive during heat stress [5, 9]. Zebu cattle are better adapted to hot environment than *Bos taurus*. Moreover, *B. indicus* has lower body metabolic rates than *B. taurus*. The zebu cattle also have a greater ability to produce sweat. The physiological adaptations to heat stress are due to characteristic hump and presence of appendages, localized fat deposition, and a light coat colour. Zebu cattle have a higher density of larger and more superficially located sweat glands. Increases in sweat production allows zebu to dissipate more heat through sweating than *B. taurus*. Moreover, the body coat of zebu cattle tends to be lighter in colour, sleeker, and shinier in contrast to the darker, denser, and typically woolly coating of European cattle. Skin appendages are also an important and recognizable feature of *B. indicus* cattle. The main function of appendages is to increase the surface area of the skin thereby lowering the metabolic rate. A very visible characteristic of all Zebu breeds is the cervo-thoracic hump. In *B. indicus*, hump fat allows decreased body fat deposition throughout the body, resulting in increased heat dissipation. The hump is an additional appendage that contributes to the increase of body surface area. Zebu cattle utilize a series of mechanisms at the cellular and molecular level resulting in a greater adaptation to high environmental temperatures. *B. indicus* is adapted to high ambient temperatures that make them the cattle type of choice for most of the extensive production systems in the tropics and subtropics. They are known for having higher tolerance to tick infestations and tick-borne diseases and also have greater capacity to digest fodders with high dietary fibre content in comparison to *B. taurus*.

Dry climates include deserts or steppe like climate. This is mostly observed in large areas of Africa, the Middle East, or Central Asia. The dry climate is characterized by seasonal rain or prolonged drought. Small ruminants are the choice of species in this region and sustain the livelihood of the people. The Awassi breed originating from several Middle Eastern countries was improved in Israel and dissipated to the rest of the world as a selected dairy breed [10]. Fat tailed sheep are characterized by a deposition of fat at the level of the hind quarters. Adults are tall and slender and tend to have long legs. Fat tailed sheep are particularly adapted to dry climates with long and persistent dry seasons. In fact, adipose tissues accumulating in the tail fat are readily mobilized in case of prolonged periods of food scarcity and correspondingly tend to decrease its size during seasonal periods of weight loss. The fat tail along with the longer legs is particularly suitable for traveling long distances in search of pasture and water. Fat tailed sheep are considered to be more tolerant to diseases and parasites, and have gregarious and defensive instincts enabling them to defend the flock from predators like jackals or foxes. These traits make fat tailed sheep the ideal group of breeds for extensive production systems in desert or semi-desert areas of the world. They play a very important role in food security by supplying milk, meat, fat, hides, and fibres, dung for fertilizing or energy source in addition to social role as cash reserve, wealth, and status indicators [11].

The salivary gland of ruminants is quite different from other mammals [12]. The amount of saliva produced is major source of fluid to rumen and provides buffer medium for ruminal fermentation. The ruminant parotid saliva with a pH of 8.2 is rich in bicarbonate buffer and mineral ions. Saliva plays an important role on food and protects ruminants against external or internal milieus. Secretion and composition of saliva are mainly under autonomic nervous system control. The saliva type changes according to various food types. Animals possessing different dietary habits secrete different saliva volumes, buffer capacity, and protein composition [13]. Goats are distributed in different ecological conditions and are supposed to be more tolerant to extreme weather conditions because of their metabolic size and water conservation capacity [14]. Short-term variations in weather pattern affect the performance of goats and are exhibited by physiological and behavioural changes in goats [15–19]. Goats are highly adaptable to heat stress due to specific anatomical and physiological characteristics. The morphological characteristics that result in better adaptation are large salivary glands, a higher surface area of absorptive mucosa and the capacity to increase the volume of the foregut with high fibrous foods [19]. The anatomical structure of goats favours better regulation of body temperature during periods of extreme heat. Short hair, abundant sweat glands, and minimal subcutaneous fat allow goats to regulate body temperature in an efficient manner. The anatomical structures of goats favour better regulation of body temperature during elevated atmospheric temperature. Short hair, sweat gland, and less subcutaneous fat are major characteristics of goats which regulate body temperature in efficient manner. Therefore, goats are adaptable varied climatic region and can be manipulated for future local and global use.

Global warming is the major concern that will define livestock production systems and livestock productivity globally and will have even greater influence on selection of livestock types and breeds in the coming decades [3]. Due to global warming the disease occurrence and susceptibility will vary in tropical regions of Asia [20, 21], Australia [22], and South America [23]. Due to global increase in atmospheric temperature, the polar region will see significant alterations in rain patterns and frequencies [3]. The tropic will witness the increase in desertification. Moreover, the vast areas in Siberia and Canada may be feasible for agricultural and animal production purposes. As tropic will be more heat prone and humid, therefore it may be required to introduce smaller heat tolerant animals, able to survive hot weather and long dry periods with limited feed and water resources. The cattle production in temperate climates will also shift toward genotypes with higher tolerance to diseases and parasites, with particular relevance to tick-borne diseases.

5.2 Thermal Stress on Livestock Growth and Development

Temperature-humidity index (THI) has been used to define environmental effect as stressful and fatal condition in different livestock with respect to climatic condition [7, 24]. THI is the most prevalent index to evaluate heat stress which combines the effect of heat and humidity. This index has been developed as a weather safety index to monitor and reduce heat stress-related losses. The variation in heat stress tolerance at cellular level in response to environmental stimuli is observed between individuals in the population. Cellular tolerance to heat stress is regulated by heat shock proteins (HSP). Heat shock proteins (HSP) are released in cell in response to various environmental and oxidative stresses [25, 26]. The regulation of HSP production is critical to cell survival. HSPs enable protein stabilization and activate various regulated proteins and block apoptosis [27, 28]. The HSP acts as a molecular chaperone by binding with other cellular proteins and facilitates intracellular transport. Among the HSPs, Hsp70 has significant role in cell thermo-tolerance and animal survival [29, 30]. HSP transcription is increased by heat shock as well other stress stimuli in different tissues including bovine embryos [31]. Hsp70 concentration in blood is a reliable indicator of chronic stress in feedlot cattle [32]. There is considerable evidence that the synthesis of Hsp70 is temperature dependent [33] and thus Hsp70 responses could be considered as cellular thermometer. HSPs are classified into several families on the basis of their molecular size and amino acid sequence similarity. The most highly conserved family of HSPs is the 70 kDa family (Hsp70), which is controlled by 13 genes in human and 4 genes in bovine [34–36]. HSP70 gene is involved in cellular stress response and also in tissue-specific and housekeeping biological tasks [37]. It has been established in mouse Hsp70gene knock-out models that the cytosolic Hsp70 family members regulate cellular stress response, while other HSPs are involved in tissue-specific and housekeeping biological tasks [37]. Heat stress is regulated in two stages such as acute (short term) and chronic (long term) [38]. The acute phase includes the shock response at the cellular level and the chronic phase results in acclimation to the stressor and involves reprogramming of gene expression and metabolism [6, 39]. In ruminants there is loss in productivity as animals pass through the acute phase and manage to restore the productivity as animals undergo acclimation to the stress. The differential and tissue-specific expression of Hsp70 proteins is not well characterized in ruminants. The characterization of cellular heat stress response has been carried out in response to high temperature in laboratory condition; however, limited studies have been carried out in vivo condition. The mechanism of role of Hsp70 in thermal tolerance and adaptation to heat and cold conditions should be analysed in various animals to understand the interaction of several genes in different pathways.

Genetic differences in thermo tolerance at the physiological and cellular levels are analysed in *Bos indicus* and *Bos taurus* [5, 40, 41]. The polymorphic pattern of hsp70 gene is associated with thermo tolerance in dairy cattle [42]. Loss in livestock productivity due to heat stress has been documented in cattle [43]. The variation in Hsp70 gene expression has been positively correlated with thermo tolerance in *Drosophila melanogaster, Caenorhabditis elegans*, rodents, and humans [44]. In farm animals, the association between differential gene expression pattern and thermal tolerance has been less studied. HSPs protect cells from negative effect of heat stress by synthesizing HSP at cellular level. Apart from heat stress, the HSPs are also synthesized by the cells in response to a variety of stimuli, including oxidative, metabolic, and chemical stress [45–47]. Heat shock (41 °C) causes increase in HSP synthesis and decrease in protein synthesis. Collier et al, [7] reported the direct effect of thermal stress on cellular growth and ductal branching of bovine mammary epithelial cells (BMECs) and also showed downregulation of genes associated with protein synthesis and cellular metabolism. The protective effect of hsp70 in differential organ such as heart and kidney tissue has been established [48]. The individuals exposed to stress elicit HSP response in the cells of various organs. The hypothermia induces Hsp72 mRNA expression in the brain, heart, kidney, liver, and lungs of mice. Zulkifli et al. [49] reported that transportation stress induced Hsp70 expression in heart and kidney of goats. Higher Hsp70 mRNA level expression provides protection due to higher muscle glycogen content and may influence meat quality in ruminants [49, 50].

5.3 Climate Change and Animal Production

5.3.1 Lactation

It has been seen that the higher milk producing cows are more susceptible to thermal load. It has been observed that milk yield decreases when the THI exceeds 72. Moreover, the studies in climate controlled experiment indicated that the milk yield decreases at THI 68 [51]. Heat stress reduces milk yield both by direct or indirect (via reduced feed intake mechanism) [52]. The reduced feed intake due to heat load reduces milk yield by 35–50% [53]. It has been observed that when heat stress exceeds a given threshold, the cumulative thermal load reduces nutrient intake and thereby decline in milk production. THI increases from 68 to 78, decreases DMI by 9.6%, and milk production by 21% [2, 54]. Milk constituents are significantly affected by heat stress during summer season. Dairy breeds are more susceptible to heat stress than meat breeds. Higher milk producing animal had increased metabolic heat production and this causes more susceptibility to heat stress as compared to low milk producing animals. Heat stress causes decline in dry matter intake and feed conversion efficiency which directly affects the body condition and resulting low milk yield [55].

Milk yield in dairy goats decreased as THI index value increased, and for each 1 unit increment of THI there is a decrease of 1% in milk yield [19, 56] summarized the HS risk on milk yield in dairy goats according to THI as following: Normal (no effect on milk yield) THI < 80; Alert (modest effect on milk yield) $80 \leq$ THI < 85; Danger (severe effect on milk yield) $85 \leq$ THI < 90; and Extreme (can result in death) THI ≥ 90. Lactating goats exposed to moderate or severe (THI $= 81$ or 89) HS for 4 days lost milk yield of 3% or 13%, respectively [25]. Brown et al. [57] also reported that exposure of dairy goats to moderate HS (34 °C; THI $= 79$) for 5 weeks depresses milk yield. Furthermore, dairy goats kept under HS in a climatic chamber reduce their milk yield by 3–10% [58]. On the contrary, Hamzaoui et al. [18] found that dairy goats were able to maintain milk yield under HS conditions (31–37 °C; THI = 77 and 65, respectively). The content of protein and protein fractions in milk were reduced under HS [18].

5.3.2 Growth

The growth phase in ruminants is not severely affected by heat load. The growing animals exhibit better adaptability to heat stress as compared to dairy animals. The livestock during growth stage may tolerate heat stress due to increased surface area to mass ratio, reduced rumen heat production, and reduced overall metabolic heat production (on body weight basis). Moreover, the cattle during growth stage compensate loss due to heat stress after a short period [59].

5.3.3 Reproduction

The economic losses are caused by sustained thermal load due to reduced growth and efficiency, increased health care cost, decreased carcass value, and increased mortality. It has been observed that heat stress had negative impact on reproductive efficiency in ruminants. Heat stress leads to infertility in animals. Heat stress causes reproductive problems through a direct effect of hyperthermia on the reproductive axis. Heat stress indirectly affects the feed intake and reduces metabolic heat production leading to changes in energy balance and nutrient availability. These changes act via hypothalamic-hypophyseal axis and cause negative effects on reproductive performance (affecting reproductive cyclicity, pregnancy, and foetal development). Heat stress affects germ cell development, pregnancy establishment, maintenance of gestation or lactation performance. The conception rates decline by 20–30% during summer months in ruminants [60]. The fertility is also reduced in ruminants and also affects early embryonic deaths [60]. Due to decrease in dry matter intake during heat stress period, there is also change in circulating concentrations of several metabolic hormones. This affects ovarian follicular development, early embryonic development, and the maternal recognition of pregnancy. Heat stress also affects oestrus detection rate, mounting activity and lowers the concentration of circulating estradiol [61]. Heat stress also affects semen quality. The semen quality in ruminants is decreased and may be due to increased production of reactive oxygen species (ROS) during period of heat stress [62].

Conception rates of lactating dairy animals have been declined with increased THI more than 72–73 in cattle [63, 64] and 75 in buffalo [65]. The release of ACTH from anterior pituitary which stimulate the release of cortisol and glucocorticoids from adrenal cortex occurs during heat stress condition. The release of luteinizing hormone is also inhibited by glucocorticoids. Extreme ambient temperatures are the major constraint to animal productivity. Heat stress is the major contributing factor to low fertility of domestic animals inseminated in the summer months [66]. Heat stress has significant effect on fertility in lactating animals [65]. Non-lactating animals and animals for meat purposes are much less likely to experience infertility during heat stress. In males, heat stress affects fertility and reduces sexual desire by reducing testosterone level, sperm production and motility, and semen quality [61, 66]. Semen characteristics (ejaculate volume, semen pH, spermatic concentration, spermatic motility, sperm abnormalities) of rams and bucks are affected within days of exposure to high environmental temperature [67]. Heat stress impairs follicular and oocyte development by altering progesterone, luteinizing and follicle-stimulating hormones' secretion and dynamics during the estrous cycle.

5.3.4 Livestock Diseases

Climate change will cause increase in weather related diseases, insect infestation, vector borne diseases. Climate change will affect the population dynamics and increase in distribution of parasites; thereby increasing disease incidence and production loss. The details of diseases due to climate change are presented in Table 5.1.

5.3.5 Feed Intake and Rumen Physiology

Heat stressed animals reduce the feed intake and this may be due to survival strategy as feed digestion produces heat in ruminants. Increase in environmental temperature has direct negative effect on appetite centre of the hypothalamus to decrease feed intake [68]. Feed intake begins to decline at air temperatures of 25–26 °C in lactating cows and reduces more rapidly above 30 °C in temperate climatic condition and at 40 °C it may decline by as much as 40% [69], 22–35% in dairy goats [18], or 8–10% in buffalo heifers [70]. Reducing feed intake is a way to decrease heat production in warm environments as the heat increment of feeding is an important source of heat production in ruminants [71]. Therefore, the animals experience a stage of negative energy balance (NEB), consequently bodyweight and body condition score goes down [72]. Increase environmental temperature alters the basic physiological mechanisms of rumen which negatively affects the ruminants with increased risk of metabolic disorders and health problems [73, 74]. Nonaka et al. [75] reported animal under HS has reduced acetate production whereas propionate and butyrate production increased as rumen function altered. Thus animal consumed less roughages changes rumen microbial population and pH from 5.82 to 6.03 [76], decreasing rumen motility and rumination [73, 74]. Subsequently, the animal health is affected due to lower saliva production, variation in digestion patterns, and decrease in dry matter intake (DMI) [73, 74]. Moreover, HS also results into hypofunction of the thyroid gland and affects the metabolism patterns of the animal to reduce metabolic heat production [77].

Table 5.1 Common livestock diseases due to climate change

Sl. no.	Type of disease
1	Vector borne diseases[a]
2	Bluetongue virus
3	Rift valley fever
4	West Nile virus
5	African horse sickness
6	Lumpy skin disease
7	Leishmaniasis
8	Epizootic haemorrhagic disease
9	Tick-borne disease
10	Parasitic diseases (excluding tick-borne)
11	Pasteurellosis
12	Avian influenza
13	Anthrax
14	Blackleg
15	Rabies
16	Tuberculosis
Sl. no.	**VBD**
1	African swine fever
2	Crimean Congo-hemorrhagic
3	Equine encephalomyelitis (eastern and western)
4	Equine infectious anaemia
5	Japanese encephalitis
6	Nairobi sheep disease
7	Vesicular stomatitis
8	Bovine anaplasmosis
9	Bovine babesiosis
10	Tularemia
11	Equine piroplasmosis
12	Heartwater
13	Leishmaniosis
14	Surra (*Trypanosoma evansi*)
15	Theileriosis
16	Trypanosomosis

[a]List of Vector borne diseases (VBD) of animals

5.3.6 Oxidative Stress

Oxidative stress results to increase in reactive oxygen species (ROS) in different cells and tissues of heat stress animals that have negative impacts on normal physiology and body metabolism. However, body has antioxidants in the form of enzymatic (super oxide dismutase [SOD], glutathione (GSH) peroxidase, and catalase) and non-enzymatic (albumin, L-cysteine, homocysteine, melatonin, and protein sulfhydryl groups). Similarly, the non-enzymatic low molecular weight antioxidants (ascorbic acid, GSH, uric acid, α-tocopherol, β-carotene, pyruvate, and retinol) provide protection against negative effects of ROS. Higher levels of catalase, SOD, GSH reductase, and malondialdehyde were observed in lactating and non-lactating buffaloes [78] and cattle [79] during summer compared to spring seasons.

5.3.7 Immunity

Thermal environment (heat or cold) affects the immune system by complex mechanisms, which includes body temperature changes, feed intake variations, behavioural and hormonal adaptation, circulatory adjustments, alteration of the acid–base balance and oxidative stress. The immune system is the major body defense systems to protect and cope against environmental stressors. Primary indicators of immunity response include white blood cells (WBCs), red blood cells (RBCs), haemoglobin (Hb), packed cell volume (PCV), glucose and protein concentration in blood get altered due to thermal stress. The blood glucose significantly decreased in HS dairy cows in accordance to greater blood insulin activity [53, 80]. Release of plasma cortisol increases in stressed animals which causes downregulation or suppression of L-selectin expression on the neutrophils surface [81]. Further, this poor L-selectin expression responsible for weak neutrophils function by failing to move into the tissue being invaded by pathogens and resulted clinical outcome of disease following exposure to an infective organism [82]. The circulating cortisol also causes increase cellular levels heat shock proteins (HSPs) that function as a danger signal to the immune system to encourage increased killing of pathogenic bacteria by neutrophils and macrophages against invading bacteria [83].

5.4 Mitigation of Environmental Stress

Livestock show various changes at behavioural, physiological, molecular/cellular, haematological, biochemical, and immunological levels in response to heat stress/ environmental stress. Environmental stress adversely affects animals' comfort, water consumption, feed intake, milk yield and quality, meat quality, and reproduction and fertility. Therefore, it is required to devise management strategies to counter the effect of climatic variation in ruminants. Adaptation strategies can improve the resilience of livestock productivity to climate change [84]. Mitigation measures could significantly reduce the impact of livestock on climate change [85]. Adaptation measures involve production and management system modifications, breeding strategies, institutional and policy changes, science and technology advances, and changing farmers' perception and adaptive capacity [5, 86, 87].

It is required to modify the production system for livestock for efficient production. Mixed cropping system was more stable and productive during climate change.

It is necessary to devise various mixed livestock and crop production, integration of livestock systems with forestry and crop production, and changing the timing and locations of farm operations [86]. Diversification of livestock and crop varieties can increase drought and heat wave tolerance. Diversity of crops and livestock is effective in fighting against climate change-related diseases and pest outbreaks [86, 88, 89]. Agroforestry (establishing trees alongside crops and pastures in a mix) as a land management approach can help maintain the balance between agricultural production, environmental protection, and carbon sequestration to offset emissions from the sector. Agroforestry may increase productivity and improve quality of air, soil, and water, biodiversity, pests, and diseases, and improve nutrient cycling [90, 91]. Changes in mixed crop–livestock systems are an adaptation measure that could improve food security [92, 93]. This type of agricultural system is already in practice in two-thirds of world, producing more than half of the milk, meat, and crops such as cereal, rice, and sorghum [94]. Changes in mixed crop–livestock systems can improve efficiency by producing more food on less land using fewer resources, such as water [94, 95].

Improving feeding practices as an adaptation measure could indirectly improve the efficiency of livestock production [96]. Some of the suggested feeding practices include modification of diets composition, changing feeding time and/or frequency [97]. Development of agroforestry and production and conservation of feed for different agro-ecological zones are adequate measure to improve the efficiency of production system. These practices can reduce the risk from climate change by promoting higher intake or compensating low feed consumption, reducing excessive heat load [97], decreasing the feed insecurity during dry seasons [98, 99], and reducing animal malnutrition and mortality [86]. Shifting locations of livestock and crop production could reduce soil erosion and improve moisture and nutrient retention [89]. Another adaptive measure could be adjusting crop rotations and changing timing of management operations (e.g. grazing, planting, spraying, irrigating). This measure can be adapted to changes in duration of growing seasons, heat waves, and precipitation variability [86, 88, 89].

5.5 Water Requirements for Livestock Production

The livestock production is an efficient source of human food and major source of livelihood in smallholder system. Therefore, it is required to analyse the water usage efficiency in producing proteins for human consumption. It is also necessary to assess the efficiency of water use in livestock with respect to dietary utilizable proteins for human consumption. Water is the critical component of the cell and accounts for more than 98% of all molecules in the body. Water is essential for the adjustment of body temperature, growth, reproduction, lactation mechanisms, digestion pattern, nutrient exchanges and transport to and from cells in blood, excretion of waste products, and heat balance. Water requirements are regulated by dry matter intake as animals require more water during drought as they are forced to select more fibrous and less digestible feed. Similarly water requirement also varies due to rise in environmental temperature (animals use more water for evaporative cooling in hot weather), and loss of water from body due to evaporation, urine, faeces, and milk. If the body loses 10% of the body water, then it becomes critical to maintain body function. Water consumption is influenced by several factors such as age, body weight gain, pregnancy, lactation, feed intake, and environmental temperature. The water requirement is fulfilled by water from external source and moisture in food stuffs.

It is now essential to use water optimally and manage the water requirement in various geographical locations for sustainable livestock production. Water is the vital component of livestock production. The average quantity of water required to produce 1 kg protein (adjusted for biological value) in various produce in different countries is presented in Table 5.2. Water usage for livestock production should be considered as an integral part of agricultural water resource management with respect to production system such as mixed cropping-livestock or grain feed system. Water requirement also varies with respect to scale of production in intensive or extensive management system. The water demands of livestock will compete with water required for crop production and water requirement of human population. Crop sector can utilize harvested rain water or stored water; however, livestock sector requires clean water. Therefore, it is necessary to analyse water requirement and water use efficiency in different livestock production system.

5.5.1 Water Requirements for Livestock

The livestock sector uses more than 8% of global water use and the major portion (7% of the global usage) of water is used for feed crop production. Water used for product, processing, drinking, and servicing livestock is very less (1% of global water). Livestock spend between 10 and 15 min per day for drinking water [101]. Water consumption for any animal can be predicted. The total water requirement can be predicted from knowledge of free water intake, body weight, dry matter content, crude protein content, and time of the year [102]. Catabolism of fat, carbohydrate, and protein produces 1190 g, 560 g, and 450 g of water/kg, respectively and these metabolic waters are important sources for all animals.

The average daily water consumption for various classes of livestock is presented in Table 5.3. Water quality is important as it affects the total water consumption by livestock as well as the health of the individual. The water with objectionable taste and odour will reduce individual water consumption, reduce livestock feed intake and weight gain. Water consumption is dependent on the animal's age, physiological condition, diet, and several other factors.

It is essential to determine the characteristics of good quality water for livestock. The water quality depends on the following parameters such as sodium (Na), alkalinity, sulfate (SO4), nitrate (NO3-N), electrical conductivity (E.C.), and hardness. All these factors need to be analysed to decide the water quality and water should be considered to be fit for consumption.

Table 5.2 Litres of water used to produce 1 kg of protein (adjusted for biological value) from 11 countries, and the average quantity of water required to produce those products along with the least significant difference values ($p = 0.05$)

	Virtual water content of products (l/kg biologically available protein)												
Product	USA	China	India	Russia	Indonesia	Australia	Brazil	Japan	Mexico	Italy	Netherlands	Average	LSD
Soybean	6756	9460	14,907	14,217	7338	7613	3890	8408	11,484	5440		8952	25,173
Wheat	15,258	12,401	29,726	42,684		28,540	29,043	13,191	19,158	43,510	11,125	24,464	25,173
Maize	14,968	24,514	59,290	42,761	39,333	22,773	36,119	45,699	53,382	16,223	12,489	33,414	24,002
Rice	25,417	26,334	56,815	47,864	42,860	20,375	61,440	24,341	43,498	33,471		38,241	25,173
Sorghum	10,577	11,673	54,821	32,219		14,622	21,763		16,393	7872		21,242	28,144
Eggs (whole)	14,650	34,443	73,067	47,725	52,392	17,891	32,376	18,279	41,496	13,476	13,622	32,674	24,002
Milk	25,703	36,982	50,629	19,741	42,271	33,839	37,019	30,030	88,092	31,842	23,706	40,896	24,002
Chicken meat	23,817	36,409	77,125	57,455	55,321	29,051	39,011	29,679	49,978	21,913	22,152	40,174	24,002
Goat meat	30,726	39,819	51,712	52,739	45,292	38,273	41,623	25,522	102,208	41,673	27,825	45,219	24,002
Sheep meat	59,588	51,862	66,717	75,978	59,379	69,259	62,479	35,601	168,267	75,490	52,819	70,676	24,002
Pork	54,974	30,802	61,257	96,782	54,862	82,321	67,122	69,128	91,376	88,841	52,800	68,206	24,002
Beef	131,529	125,218	164,319	209,641	147,729	170,600	169,094	109,855	376,472	211,026	114,455	175,631	24,002

Adapted from Hoekstra and Chapagain [100]

Table 5.3 Water Consumption for various classes of livestock

Species	Water consumption (gallons per day)
Beef cattle	7–12 per head
Dairy cattle	10–16 per head
Horses	8–12 per head
Swine	3–5 per head
Sheep and goats	1–4 per head
Chickens	8–10 per 100 birds
Turkeys	10–15 per 100 birds

Adapted from Oslon and Fox [103]

The livestock sector is currently growing faster than other forms of agriculture in almost all countries. The animal production is shifting towards more industrialization to optimize the environmental constraints. However, the major proportion of milk, meat, and cereals in developing countries still comes from mixed crop–livestock systems. Mixed farming livestock production systems are widely distributed in both the developed and developing world and contribute significantly to production. Mixed farming systems are those in which more than 10% of the dry matter feed for animals comes from crop by-products or more than 10% of the total value of production comes from non-livestock farming activities. Mixed farming is a form of agricultural production in which crop residue wastes being used by livestock, and livestock waste being returned to the crops as fertilizer. Maximizing the water efficiency of the mixed farming system will improve the water productivity of both livestock and cropping systems [104].

This integrated system will need a multi-disciplinary approach to deal with the interactions between the plant and livestock systems and determine their interdependence. Livestock provide not only meat and milk, but also draught power for cropping and transport, fertilizer from their manure, and financial security during harsh times. The water efficiency of grazing cattle can be improved by better use of rain-fed pastures. It is generally suggested that the water points need to be developed for better utilization of grazing lands and to allow the movement of livestock across the region by the strategic placement of watering points. In the case of ruminants, better placing of watering points encourages livestock to graze the available pasture more uniformly as they prefer to graze close to their drinking water supplies.

Gerrish and Davis [105] found that 77% of cattle grazing took place within 366 m of the water source but that 65% of the available pasture in the paddock was more than 730 m from the watering point. The costs associated with including extra water points have not yet been considered but are being investigated in the extensive grazing lands of northern Australia [106, 107]. Improving access to water will have a significant impact on the virtual water content of pasture-grazed livestock, if pasture use can be economically increased from 10% to 20% in these extensive grazing lands [104].

In some developing countries, livestock production has shifted towards monogastrics, with pigs and poultry accounting for 77% of the expansion in production. Total meat production in developing countries is more than tripled between 1980 and 2004. The growth in ruminant production was only 111%, with that of monogastric production expanding by more than fourfold over the same period. It is important to note that these figures mask considerable diversity within and among countries. However, in developed countries, total livestock production is increased by only 22% between 1980 and 2004. In these countries, there was a differential growth between livestock sectors, with ruminant production declining by 7% and pig and poultry production increasing by 42% [95].

Livestock farming is likely to compete with other farming sectors for water because of increasing pressures from global climate change and the variability of water availability and distribution. The challenge for the future is to optimize livestock water productivity by improving animal productivity within the framework of integrated soil-water-animal management, under both rain-fed and irrigated conditions.

References

1. Thornton PK (2010) Livestock production: recent trends, future prospects. Philos Trans R Soc B Biol Sci 365:2853–2867
2. Spiers DE, Spain JN, Sampson JD et al (2004) Use of physiological parameters to predict milk yield and feed intake in heat-stressed dairy cows. J Therm Biol 29:759–764
3. Miraglia M, Marvin HJ, Kleter GA et al (2009) Climate change and food safety: an emerging issue with special focus on Europe. Food Chem Toxicol 47:1009–1021
4. Mirkena T, Duguma G, Haile A et al (2010) Genetics of adaptation in domestic farm animals: a review. Livest Sci 132:1–12
5. Hansen PJ (2004) Physiological and cellular adaptations of zebu cattle to thermal stress. Anim Reprod Sci 82–83:349–360
6. Collier RJ, Collier JL, Rhoads RP et al (2008) Invited review: genes involved in the bovine heat stress response. J Dairy Sci 91:445–454
7. Collier RJ, Stiening CM, Pollard BC et al (2006) Use of gene expression microarrays for evaluating environmental stress tolerance at the cellular level in cattle. J Anim Sci 84 (E. Suppl.):E1–E13
8. Barros CM, Pegorer MF, Vasconcelos JL et al (2006) Importance of sperm genotype (indicus versus taurus) for fertility and embryonic development at elevated temperatures. Theriogenology 65:210–218
9. Berman A (2011) Invited review: are adaptations present to support dairy cattle productivity in warm climates? J Dairy Sci 94(5):2147–2158
10. Gootwine E (2011) Mini review: breeding Awassi and Assaf sheep for diverse management conditions. Trop Anim Health Prod 43:1289–1296
11. Udo HM, Budisatria IG (2011) Fat-tailed sheep in Indonesia; an essential resource for smallholders. Trop Anim Health Prod 43:1411–1418
12. Steward MC, Poronnik P, Cook DI (1996) Bicarbonate transport in sheep parotid secretory cells. J Physiol 494:819–830
13. Sales-Bapista E, Lamy E, Mau M et al (2009) Variation in salivary protein composition related to feeding behavior and its ecological implications. In: Zhang W, Liu H (eds) Behavioural and chemical ecology. Nova Science Publisher, Hauppauge
14. Silanikove N (2000) Effects of heat stress on the welfare of extensively managed domestic ruminants. Livest Prod Sci 67:1–18
15. Appleman RD, Delouche JC (1958) Behavioral, physiological and biochemical responses of goats to temperature, 0° to 40° C. J Anim Sci 17(2):326–335

16. Solanki GS (1989) Physiological response of goat under short term exposure of low ambient temperature in marginal ad semi-arid region at Rajkot. Proc Indian Natl Sci Acad B55 5 & 6:361–368
17. Kannan G, Terrill TH, Kouakou B et al (2000) Transportation of goats: effects on physiological stress responses and live weight loss. J Anim Sci 78(6):1450–1457
18. Hamzaoui S, Salama AAK, Caja G et al (2012) Milk production losses in early lactating dairy goats under heat stress. J Dairy Sci 95(2):672–673
19. Silanikove N, Koluman N (2015) Impact of climate change on the dairy industry in temperate zones: predications on the overall negative impact and on the positive role of dairy goats in adaptation to earth warming. Small Rumin Res 123:27–34
20. Van den Bossche P, Coetzer JA (2008) Climate change and animal health in Africa. Rev Sci Tech 27(2):551–562
21. Forman S, Hungerford N, Yamakawa M et al (2008) Climate change impacts and risks for animal health in Asia. Rev Sci Tech 27(2):581–597
22. Black PF, Murray JG, Nunn MJ (2008) Managing animal disease risk in Australia: the impact of climate change. Rev Sci Tech 27(2):563–580
23. Pinto J, Bonacic C, Hamilton-West C et al (2008) Climate change and animal diseases in South America. Rev Sci Tech 27(2):599–613
24. McDowell RE, Hooven NW, Camoens JK (1976) Effects of climate on performance of Holsteins in first lactation. J Dairy Sci 5:965–971
25. Sonna LA, Fujita J, Gaffin SL et al (1985) Invited review: effects of heat and cold stress on mammalian gene expression. J Appl Physiol 92(4):1725–1742
26. Hecker JG, Sundram H, Zou S et al (2008) Heat shock proteins HSP70 and HSP27 in the cerebral spinal fluid of patients undergoing thoracic aneurysm repair correlate with the probability of postoperative paralysis. Cell Stress Chaperones 13(4):435–444
27. Chirico WJ, Waters MG, Blobel G (1998) 70K heat shock related proteins stimulate protein translocation into microsomes. Nature 332:805–810
28. Kregel KC (2002) Heat shock proteins: modifying factors in physiological stress responses and acquired thermo tolerance. J Appl Physiol 92:2177–2186
29. Barbe MF, Tytell M, Gower DJ et al (1998) Hyperthermia protects against light damage in the rat retina. Science 241:1817–1820
30. King YT, Lin CS, Lin JH et al (2002) Whole-body hyperthermia induced thermotolerance is associated with the induction of heat shock protein 70 in mice. J Exp Biol 205:273–278
31. Wrenzycki C, Wells D, Herrmann D et al (2001) Nuclear transfer protocol affects messenger RNA expression patterns in cloned bovine blastocysts. Biol Reprod 65:309–317
32. Gaughan JB, Bonner SL, Loxton I et al (2013) Effects of chronic heat stress on plasma concentration of secreted heat shock protein 70 in growing feedlot cattle. J Anim Sci 91 (1):120–129
33. Zulkifi I, Liew PK, Israf DA et al (2003) Effect of early age feed restriction and thermal conditioning on heterophil/lymphocyte ratio, heat shock 70 and body temperature of male broiler chickens subjected to acute heat stress. J Therm Biol 28:217–222
34. Grosz MD, Womack JE, Skow LC (1992) Syntenic conservation of HSP70 genes in cattle and humans. Genomics 14:863–868
35. Gallagher DS, Grosz MD, Womack JE et al (1993) Chromosomal localization of HSP70 genes in cattle. Mamm Genome 4:388–390
36. Kampinga HH, Hageman J, Vos MJ et al (2009) Guidelines for the nomenclature of the human heat shock proteins. Cell Stress Chaperones 14:105–111
37. Daugaard M, Rohde M, Jäättelä M (2007) The heat shock protein 70 family: highly homologous proteins with overlapping and distinct functions. FEBS Lett 581:3702–3710
38. Garret AT, Goosen NG, Rehrer NG et al (2009) Induction and decay of short-term heat acclimation. Eur J Appl Physiol 107:659–670
39. Horowitz M (2002) From molecular and cellular to integrative heat defense during exposure to chronic heat. Comp Biochem Physiol A Mol Integr Physiol 131:475–483

40. Paula-Lopes FF, Chase CC Jr, Al-Katanani YM et al (2003) Genetic divergence in cellular resistance to heat shock in cattle: differences between breeds developed in temperate versus hot climates in responses of pre-implantation embryos, reproductive tract tissues and lymphocytes to increased culture temperatures. Reproduction 125(2):285–294
41. Lacetera N, Bernabucci U, Scalia D et al (2006) Heat stress elicits different responses in peripheral blood mononuclear cells from Brown Swiss and Holstein cows. J Dairy Sci 89:4606–4612
42. Basirico L, Morera P, Primi V et al (2011) Cellular thermotolerance is associated with heat shock protein 70.1 genetic polymorphisms in Holstein lactating cows. Cell Stress Chaperones 16(4):441–448
43. Bernabucci U, Lacetera N, Baumgard LH et al (2010) Metabolic and hormonal adaptations to heat stress in domesticated ruminants. Animal 4:1167–1183
44. Hashmi G, Hashmi S, Selvan S et al (1997) Polymorphism in heat shock protein gene (Hsp70) in entomopathogenic nematodes (rhabditida). J Therm Biol 22:143–149
45. Welch WJ (1992) Mammalian stress response: cell physiology, structure/function of stress proteins, and implications for medicine and disease. Physiol Rev 72:1063–1081
46. Morimoto RI, Tissieres A, Georgopoulos C (1994) Progress and perspectives on the biology of heat shock proteins and molecular chaperones. In: Morimoto RI, Tissieres A, Georgopoulos C (eds) The biology of heat shock proteins and molecular chaperones. Cold Spring Harbour Laboratory Press, Cold Spring Harbour, pp 1–30
47. Airaksinen S, Jokilehto T, Rabergh CMI et al (2003) Heat- and cold-inducible regulation of HSP70 expression in zebrafish ZF4. Comp Biochem Physiol 36:275–282
48. Latchman DS (2001) Heat shock protein and cardiac protection. Cardiovasc Res 51:637–646
49. Zulkifli I, Norbaiyah B, Cheah YW et al (2010) A note on heat shock protein 70 expression in goats subjected to road transportation hot, humid tropical conditions. Animal 4:973–976
50. Zulkifli I, Al-Aqil A, Omar AR et al (2008) Housing system affects stress and fear reaction and meat quality of broiler chickens subjected to road transportation. In: Proceeding of XXIII world's poultry congress. World's Poultry Science Association, Odijk, p 380
51. Zimbleman RB, Rhoads RP, Baumgard LH et al (2009) Revised temperature humidity index (THI) for high producing dairy cows. J Dairy Sci 92(E-Suppl 1):347
52. Baumgard LH, Rhoads RP (2012) Effects of environment on metabolism. In: Collier RJ, Collier JL (eds) Environmental physiology of livestock. Wiley, Inc., Ames, pp 81–100
53. Rhoads ML, Rhoads RP et al (2009) Effects of heat stress and plane of nutrition on lactating Holstein cows: I. production, metabolism and aspects of circulating somatotropin. J Dairy Sci 92:1986–1997
54. Bouraoui R, Lahmar M, Majdoub A et al (2002) The relationship of temperature-humidity index with milk production of dairy cows in a Mediterranean climate. Anim Res 51 (6):479–491
55. Wilson SJ, Marion RS et al (1998) Effects of controlled heat stress on ovarian function of dairy cattle. J Dairy Sci 81:2124–2131
56. Salama AAK, Caja G, Hamzaoui S et al (2014) Different levels of response to heat stress in dairy goats. Small Rumin Res 121(1):73–77
57. Brown DL, Morrison SR, Bradford GE (1988) Effects of ambient temperature on milk production of Nubian and Alpine goats. J Dairy Sci 71:2486–2490
58. Salama AAK, Hamzaoui S, Badaoui B et al (2012) Transcriptome analysis of blood in heat-stressed dairy goats. J Dairy Sci 95(Suppl. 2):188
59. Mitlöhner FM, Morrow JL et al (2001) Shade and water misting effects on behavior, physiology, performance, and carcass traits of heat-stressed feedlot cattle. J Anim Sci 79:2327–2335
60. Rensis FD, Scaramuzzi RJ (2003) Heat stress and seasonal effects on reproduction in the dairy cow-a review. Theriogenology 60:1139–1151
61. Hansen PJ, Aréchiga CF (1999) Strategies for managing reproduction in the heat-stressed dairy cow. J Anim Sci 77(Suppl 2):36–50

62. Nichi M, Bols PE, Züge RM et al (2006) Seasonal variation in semen quality in Bos indicus and Bos taurus bulls raised under tropical conditions. Theriogenology 66:822–828
63. Morton JM, Tranter WP, Mayer DG et al (2007) Effect of environmental heat on conception rates in lactating dairy cows: critical periods of exposure. J Dairy Sci 90:2271–2278
64. Schuller LK, Burfeind O, Heuwieser W (2014) Impact of heat stress on conception rate of dairy cows in the moderate climate considering different temperature humidity index thresholds, periods relative to breeding, and heat load indices. Theriogenology 81:1050–1057
65. Dash S (2013) Genetic evaluation of fertility traits in relation to heat stress in Murrah buffaloes. M.V.Sc. Thesis, ICAR-NDRI (Deemed University), Karnal
66. Hansen PJ, Drost RM, Rivera FF et al (2001) Adverse impact of heat stress on embryo production: causes and strategies for mitigation. Theriogenology 55:91–103
67. Gimenez D, Rodning S (2007) Reproductive management of sheep and goats. Alabama Cooperative Extention. System (ACES), http://www.aces.edu/pubs/docs/A/ANR1316/ANR-1316.pd. Accessed 16 June 2014
68. Baile CA, Forbes JM (1974) Control of feed intake and regulation of energy balance in ruminants. Physiol Rev 54(1):160
69. Rhoads RP, Baumgard LH, Suagee JK et al (2013) Nutritional interventions to alleviate the negative consequences of heat stress. Adv Nutr 4(3):267–276
70. Hooda OK, Singh S (2010) Effect of thermal stress on feed intake, plasma enzymes and blood bio-chemicals in buffalo heifers. Indian J Anim Nutr 27(2):122–127
71. Kadzere CT, Murphy MR, Silanikove N et al (2002) Heat stress in lactating dairy cows: a review. Livest Prod Sci 77(1):59–91
72. Lacetera N, Bernabucci U, Ronchi B et al (1996) Body condition score, metabolic status and milk production of early lactating dairy cows exposed to warm environment. Riv Agric Subtrop Trop 90(1):43–55
73. Nardone A, Ronchi B, Lacetera N et al (2010) Effect of climate changes on animal production and sustainability of livestock systems. Livest Sci 130(1-3):57–69
74. Soriani N, Panella G, Calamari L (2013) Rumination time during the summer season and its relationships with metabolic conditions and milk production. J Dairy Sci 96(8):5082–5094
75. Nonaka I, Takusari N, Tajima K et al (2008) Effects of high environmental temperatures on physiological and nutritional status of pre-pubertal Holstein heifers. Livest Sci 113(1):14–23
76. Hall MB (2009) Heat stress alters ruminal fermentation and digesta characteristics, and behavior in lactating dairy cattle. In: Chilliard Y, Glasser F, Faulconnier Y, Bocquier F, Veissier I, Doreau M (eds) Proceeding of 11th international symposium on ruminant physiology. Wageningen Academic Publication, Wageningen, p 204
77. Helal A, Hashem ALS, Abdel-Fattah MS et al (2010) Effect of heat stress on coat characteristics and physiological responses of Balady and Damascus goats in Sinai, Egypt. Am Eurasian J Agric Environ Sci 7(1):60–69
78. Lallawmkimi MC, Singh SV, Upadhyay RC et al (2013) Impact of vitamin supplementation on heat shock protein 72 and antioxidant enzymes in different stages of Murrah buffaloes during seasonal stress. Indian J Anim Sci 83(9):909–915
79. Yatoo MI, Dimri M, Sharma MC (2014) Seasonal changes in certain blood antioxidants in cattle and buffaloes. Indian J Anim Sci 84(2):173–176
80. Baumgard LH, Rhoads RP (2013) Effects of heat stress on post absorptive metabolism and energetics. Annu Rev Anim Biosci 1:311–337
81. Burton JL, Ronald JE (2003) Immunity and mastitis: some new ideas for an old disease. Vet Clin Food Anim Pract 19(1):1–45
82. Kansas GS (1996) Selectins and their ligands: current concepts and controversies. Blood 88 (9):3259–3287
83. Do Amaral BC, Connor EE, Tao S et al (2011) Heat stress abatement during the dry period influences metabolic gene expression and improves immune status in the transition period of dairy cows. J Dairy Sci 94(1):86–96

84. USDA (United States Department of Agriculture) (2013) Climate change and agriculture in the United States: effects and adaptation. USDA Technical Bulletin, Washington, DC. http://www.usda.gov/oce/climate_change/effects_2012/CC%20and%20Agriculture%20Report%20%2802-04-2013%29b.pdf
85. Dickie A, Streck C, Roe S et al (2014) Strategies for mitigating climate change in agriculture: abridged report. Climate focus and California environmental associates, prepared with the support of the climate and land use Alliance. Report and supplementary materials. www.agriculturalmitigation.org
86. IFAD (International Fund for Agricultural Development) (2010) Livestock and climate change. http://www.ifad.org/lrkm/events/cops/papers/climate.pdf
87. Rowlinson P, Steele M, Nefzaoui A (2008) Livestock and global climate change: adaptation I and II. In: Rowlinson P, Steel M, Nefzaoui A (eds) Livestock and global climate change conference proceeding. Cambridge University Press, Tunisia, pp 56–85
88. Batima P, Bat B, Tserendash L (2005) Adaptation to climate change, vol 90. ADMON Publishing, Ulaanbaatar
89. Kurukulasuriya P, Rosenthal S (2003) Climate change and agriculture: a review of impacts and adaptations. Climate Change Series Paper No. 91, World Bank, Washington DC
90. Jose S (2009) Agroforestry for ecosystem services and environmental benefits: an overview. Agrofor Syst 76:1–10
91. Smith JP, Pearce BD, Wolfe MS (2012) Reconciling productivity with protection of the environment: is temperate agroforestry the answer? Renewable Agric Food Syst 28(1):80–92
92. Herrero M, Thornton PK, Notenbaert AM et al (2010) Smart investments in sustainable food production: revisiting mixed crop livestock systems. Science 327:822–825
93. Wani SP, Rockstrom J, Oweis T (2009) Rainfed agriculture: unlocking the potential. CAB International, Wallingford
94. Herrero M, Thornton PK, Notenbaert A et al (2012) Drivers of change in crop–livestock systems and their potential impacts on agro-ecosystems services and human wellbeing to 2030: a study commissioned by the CGIAR system wide livestock Programme. International Livestock Research Institute, Nairobi
95. Steinfeld H, Gerber P, Wassenaar T et al (2006) Livestock's long shadow: environmental issues and options. FAO, Rome
96. Havlík P, Valin H, Mosnier A et al (2013) Crop productivity and the global livestock sector: implications for land use change and greenhouse gas emissions. Am J Agric Econ 95:442–448
97. Renaudeau D, Collin A, Yahav S et al (2012) Adaptation to hot climate and strategies to alleviate heat stress in livestock production. Animal 6(05):707–728
98. Thornton PK, Herrero M (2010) The inter-linkages between rapid growth in livestock production, climate change, and the impacts on water resources, land use, and deforestation. World Bank Policy Research Working Paper, WPS 5178. World Bank, Washington, DC
99. Thornton PK, Herrero M (2010) Potential for reduced methane and carbon dioxide emissions from livestock and pasture management in the tropics. Proc Natl Acad Sci U S A 107(46):19667–19672
100. Hoekstra AY, Chapagain AK (2007) Water footprints of nations: water use by people as a function of their consumption pattern. Water Resour Manag 21:35–48
101. Meyer U, Everinghoff M, Gädeken D, Flachowsky G (2004) Investigations on the water intake of lactating dairy cows. Livest Prod Sci 90(2-3):117–121
102. Holter JB, Urban WF Jr (1992) Water partitioning and intake prediction in dry and lactating Holstein cows. J Dairy Sci 75(6):1472–1479
103. Olson OE, Fox DG (1981) Great plains beef cattle feeding handbook. GPE-1401. South Dakota State University. Brookings
104. Peden D, Tadesse G, Misra AK et al (2007) Water and livestock for human development. In: Molden D (ed) Water for food, water for life: comprehensive assessment of water management in agriculture. Earthscan/International Water Management Institute, London/Colombo, 645 pp. www.ilri.org/data/livelihood/cpwf/Chapter%2013.pdf. Accessed 6 Dec 2007

105. Gerrish J, Davis M (1999) Water availability and distribution. In: Gerrish J, Roberts C (eds) Missouri grazing manual (eds) University of Missouri Extension, Columbia
106. Nagy KA (1994) Seasonal water, energy and food use by free-living, arid-habitat mammals. Aust J Zool 42:55–63
107. National Research Council (1981) Effect of environment on nutrient requirements of domestic animals. National Academy Press, Washington, DC. www.nap.edu/openbook.php?record_id=4963&page=R1. Accessed 4 Dec 2007

Sustainable Livestock Farming

6

The practice of farming, breeding, and slaughter of livestock is known as animal husbandry, which has been practiced worldwide for food and commodities. In due time, the livestock farming practices have been shifted to intensive animal farming, often referred as "Factory farming". Intensive livestock farming produces various commercial products, but has also led to negative impact on animal welfare, the environment, and public health. Livestock fulfils the nutritional requirement of human being. The livelihoods of the people living in poverty worldwide depend directly on livestock. Livestock provides 18% of the total calories and 39% of the protein consumption. The annual growth rates of livestock sector have been around 3.8%, as compared to 2.7% for crops and 1.2% for non-food agricultural products. The overall growth in the livestock has consistently exceeded that of the crop sector. But the growth rates of livestock only getting slow down over the projection period (Table 6.1). The livestock sector consumes 6 billion tonnes of dry feed per year. This includes one-third of global cereal production. About 86% of global livestock feed intake is made of materials that are not consumable by human beings. About a third of total value of gross agricultural output is consumed by the developing countries, and it is apprehended to rise quickly due to increase in population, process of urbanization, rising consumer incomes, and potential dietary shifts to higher calorie intake, and use of greater amount of animal proteins [1].

So, the livestock sector has been emerging as great value for the sustainability of rural economic of developing nations, in spite of its much adverse impact on natural ecosystem. Due to the growing demand for animal proteins, it is essential to design new model on livestock production systems, ensuring food security and sustainability. Sustainable farming means conservation of manmade ecosystem for any agricultural practice like development of manmade multi crops cultivation, including the practice of fisheries. For a farm to be sustainable, it must produce sufficient quantity of nutrient rich foods, be environmentally safe with return guarantee, and sustainable in nature. In a sustainable farming practice the farmer should minimize the external input like chemical fertilizers and non-conventional energy, and rely, as much as possible on the renewable resources available in and

P. K. Rout, B. K. Behera, *Sustainability in Ruminant Livestock*,
https://doi.org/10.1007/978-981-33-4343-6_6

Table 6.1 The annual growth rates of total livestock production

	Year					
	1969–1999	1979–1999	1989–1999	1997/1999–2015	2015–2030	1997/1999–2030
World	2.2	2.1	2.0	1.7	1.5	1.6
Excl. China	1.7	1.3	0.8	1.6	1.5	1.5
Developing countries	4.6	5.0	5.5	2.6	2.1	2.4
Excl. China	3.5	3.5	3.6	2.8	2.5	2.7
Excl. China and Brazil	3.3	3.3	3.3	2.9	2.6	2.8
Sub-Saharan Africa	2.4	2.0	2.1	3.2	3.3	3.2
Latin America and the Caribbean	3.1	3.0	3.7	2.4	1.9	2.1
Excl. Brazil	2.3	2.1	2.7	2.4	2.1	2.3
Near east/North Africa	3.4	3.4	3.4	2.9	2.6	2.7
South Asia	4.2	4.5	4.1	3.3	2.8	3.1
East Asia	7.2	8.0	8.2	2.3	1.6	2.0
Excl. China	4.8	4.7	3.7	3.0	2.7	2.8
Industrial countries	1.2	1.0	1.2	0.7	0.4	0.6
Transition countries	−0.1	−1.8	−5.7	0.5	0.6	0.5

Total livestock production was derived by aggregating four meats, milk and eggs at 1989/1991 international commodity prices used to construct the FAO indices of agricultural production (www.fao.org)

around the farming system. This sort of practice would be immensely helpful in developing eco-friendly sustainable farm in rural area of developing countries. In this context the idea of sustainable organic livestock farming has come to limelight. Currently, integrated sustainable livestock farming has drawn attention of farmers due to its continuous return of diversified products, besides livestock as primary target (Fig. 6.1). Sustainable integrated livestock faming is a human manipulated ecosystem. Integrated sustainable livestock farming is a complex issue associated with producing protein rich foods while maintaining natural resources including soil, water, and biota with no adverse impact on the wider environment. So, people have started to have option for organic livestock farming in order to fulfil the demand for animal proteins and other value-added products derived from livestock. For developing such organic livestock farming, it is also necessary to take care of environment protection, food safety, animal welfare, and social ethical issues.

6.1 Livestock as Driving Force for Food Security

Animals and human being have evolved in different angle and developed specialized digestive and metabolic systems to process food and absorb nutrients. Herbivores and carnivores differ in dental structure to digestive tract and enzymatic pattern for food processing and absorption. Ruminants are herbivores and digest cellulose (hay, grass, straws) via specialized digestive tract called rumen. The feed and fodder

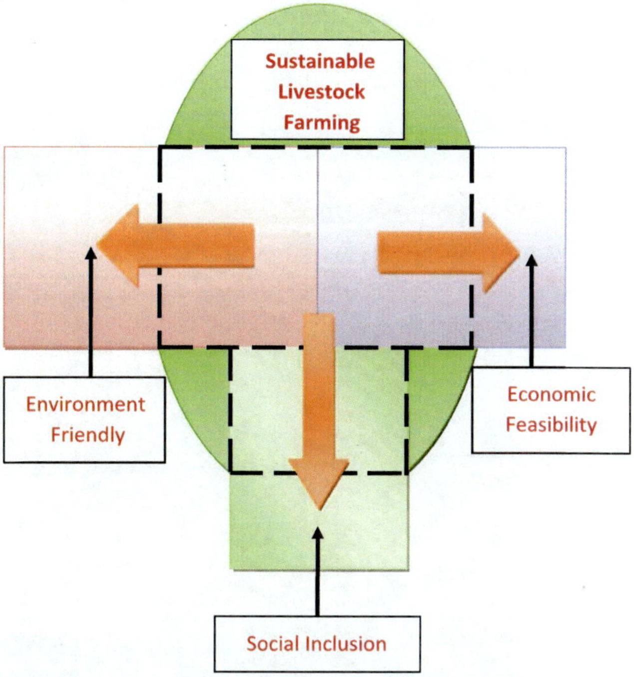

Fig. 6.1 Sustainable livestock farming, graphic model

which are not fit for human consumption can be utilized by ruminants and convert into best edible protein in nature. Monogastric animals have some limitation in using cellulose as food. Ruminant animals will be most valuable because they can convert the energy and protein in fibrous feedstuffs to milk, meat, wool, and other products. Ruminant systems mainly depend on pasture-based systems. However, monogastric animals are fed considerable quantities of human-edible grains to support their productivity. Similarly, crop and industrial by-products are used as feed in both ruminant and monogastric systems; however, it is necessary to assess the improved feed efficiency of the monogastric animal and the quantity of human-edible grains used for monogastric feed.

The FAO estimates that 26% of the world land area and 70% of the world agricultural area are covered by grasslands [2]. The grassland is used for the grazing of cattle, sheep, goats, water buffalo, and wildlife, and produces human-edible food. Ruminant animals are best equipped to harvest the solar energy stored in the fibrous feeds growing on these grasslands and convert it to meat, milk, wool, or power from draft animals [3]. The major energy stored in these plants is in cellulose or hemicelluloses, which are inefficiently digested by monogastrics, and are not digestible by man. Ruminants are unique because of the symbiotic relationship the animal has with the microflora in the rumen. The rumen is the largest of the four compartments in the stomach and houses different types of bacteria, protozoa, and fungi. The microbial population in the rumen produces a complex of enzymes that breaks down the plant cellulose and hemicelluloses into simple sugars. Because of the anaerobic environment within the rumen, the sugars are fermented into volatile

Fig. 6.2 Livestock farming and food security

fatty acids, primarily acetic, propionic, and butyric acids. These volatile fatty acids typically supply approximately 70% of the energy needed by ruminants. Ruminants have another advantage over humans and other monogastric animals that they can use non-protein nitrogen as a protein source. In many types of forage, as much as 70% of the nitrogen is bound to fibre or found in non-protein forms [4]. Rumen bacteria can convert nitrogen to microbial protein in the rumen. The microbial protein then passes out of the rumen and is broken down to amino acids in the small intestine where they are absorbed. Consequently, the two major requirements for growth and milk production, energy and protein, can be harvested from grasslands much more efficiently by ruminants than any other species.

Economic factors are strong predictors of food consumption and the dietary intake of food components. In the United States and other advanced economies, more than 60% of protein comes from animal sources, whereas the contribution of animal sources in Africa, India, and other food-deficit countries is 20–25% of total protein. The demand for animal protein is increasing over the years for food security (Fig. 6.2) as the world population is also increasing over the years. Global grain production has increased twofold during the past 50 years; however, the hunger population is also increasing since 2015. The developing regions lack the facilities and infrastructure to store and transport cereal crops. The transport infrastructure, adoption of best management practices, and development of better-functioning markets are the major factors which can prevent the post-harvest loss. There are number of underlying issues behind food shortages. The competition between animal and human for feed is not the major issues behind food shortage. The most significant contributor to food shortages in developing regions may be income distribution.

It has been observed that one-third of cereals are fed to livestock [5] to produce animal protein. The treatment of crop residues with calcium oxide improves fibre digestion and animal performance [6–8]. It has been demonstrated that combination of by-product feeds (distillers' grains) and treated straw/corn stover can replace approximately two-thirds of the corn typically fed to finishing cattle while maintaining equal performance. Feed efficiency is defined as the amount of feed required to produce a unit of weight gain, milk production, or dozen eggs. In ruminants, feed conversion efficiency is defined as the amount of dry feed required per unit of weight gain and this is the most vital parameter to determine the economic cost of production. It has long been recognized that animals require a certain amount of feed nutrients each day to fulfil the maintenance requirement. Subsequently, after fulfilling the maintenance requirement, the dietary nutrient is utilized for weight gain, pregnancy, and lactation. The efficiency of feed utilization above maintenance is often referred to as partial efficiency.

Feed: gain ratio has been shown to have antagonistic genetic correlations with mature size and maintenance requirements. Selecting a decreased feed: gain ratio (i.e. more efficient animals) will result in larger late-maturing animals. To avoid these problems, Koch et al. [9] described the concept of residual feed intakes (RFIs) as the observed feed intake minus the predicted feed intake (as calculated from a regression equation that includes animal weight and gain) [10]. Animals with a negative RFI are efficiently superior because they eat less than predicted based on their average weight and daily gain. The potential advantages of negative RFI animals are multifaceted. RFI enhances more animal protein produced per unit of feed consumed, and the consumed feed is digested and used more efficiently. It has been reported that low-RFI animals produced 16,100 L less methane per year than high-RFI animals at the same level of daily gain [11]. Feed conversions are expressed on an "as-fed basis", in which case the final diet may contain between 10% and 70 moisture. To avoid these problems, all diets should be converted to an equal dry matter basis.

Food-residue disposal is a significant issue in the countries having large food-processing industries. It is estimated that more than half of the industrial waste in the Netherlands originates from the food-processing industry, yet livestock can often convert this residue into meat, milk, and wool. Given the perceived competition between animals and humans for feed and food, the most meaningful measure of feed efficiency in the future may be the ratio of human-edible protein input relative to the human-edible protein output. Wilkinson [12] used this approach to compare different food production systems in the United Kingdom. When expressed as the ratio of human-edible protein input per unit to edible animal protein output, milk was 0.71, various beef production systems were 0.92–3.00, pork was 2.6, poultry meat was 2.1, and eggs were 2.3. The edible protein conversion ratios reported by cast (1999) were 0.48, 0.84, 3.4, and 1.6 for the milk, beef, pigs, and poultry systems in the USA, respectively.

6.2 Livestock Products

Most of the domesticated animals like cows, goats, sheep, and pigs are grown under the local environment to produce household commodities like animal proteins (meat, milk, egg) and value-added by-products.

6.2.1 Food Products from Livestock

Meat is obtained from variety of livestock reared in different farming systems. Meat from livestock, especially cattle, is an important source of nutrition for many people around the world. The demand for meat in global market is in increasing order since the last five decades. The global meat production showed fourfold increase during the last five decades. It has been estimated that about 80 billion animals are slaughtered each year for meat. Asia is the largest meat producer, accounting about 40–45% of total meat production. Meat consumption trends vary significantly across the world.

The quality of meat is affected by extinction of variety of plants and animals [13–17]. In 2019, Intergovernmental Science-Policy on Biodiversity and Ecosystem Services (IPBES) was established to improve the interface between science and policy on issues of biodiversity and ecosystem services. It has been identified that industrial agriculture and overfishing are the primary drivers of the extinction crisis process of plants and animal which ultimately affect meat and dairy industries.

World milk production entirely depends on the population of cattle, buffaloes, goats, sheep, and camels. Yaks, reindeers, and donkey are rare source of milk. The world milk production rate is in increasing rate of magnitude with the increase in population growth (Table 6.2). The quality of milk production depends on breeds of cattle and their geographic location. The key factors responsible for milk production are feed, water, and climate. Besides, market demand, dietary traditions, and the socio-economic conditions are also responsible for holding small-scale farmhouse at village level. The cattle is the major producer of milk and contributes about 81% of world milk production followed by buffaloes with 15%, and goat, sheep, and camel combined (4%).

Table 6.2 World milk production

	World: Milk production and uses (in milk equivalent)							
'000 tons	2015	2016	2017	2018	2019	2020	±% on '19	±% on '18
Milk (M)	413,129	409,879	389,649	393,865	395,474	19,065	−95.18%	−95.16%
Consumption of fluid milk (CM)	65,129	63,719	63,012	61,800	61,071	60,772	−0.49%	−1.66%
% (CM)/(M)	15.8%	15.5%	16.2%	15.7%	15.4%	318.8%		
Cheese (C)	142,167	143,934	143,381	146,253	147.383	2787	−98.11%	−98.09%
% (C)/(M)	34.4%	35.1%	36.8%	37.1%	37.3%	14.6%		
WMP (W)	20,695	19,688	20,219	20,859	21,016	250	−98.81%	−98.80%
% (W)/(M)	5.0%	4.8%	5.2%	5.3%	5.3%	1.3%		
SMP (S)	36,785	37,578	37,010	36,677	38,526	1426	−96.30%	−96.11%
% (S)/(M)	8.9%	9.2%	9.5%	9.3%	9.7%	7.5%		
CM + C + W + S	264,776	264,918	263,622	265,590	267,995	65,234		
% (CM + C + W + S)/(M)	64.1%	64.6%	67.7%	67.4%	67.8%	342.2%		

6.2.2 Value-Added Products from Livestock

Besides animal proteins, livestock farming system also gives many value-added products like wool (for textile industry), leather (for household materials like bag, water bag mainly for desert area use), animal fats (for soap industry), candle, and leather as heat insulator (Fig. 6.3). Gelatine is one of the most demand products made from animal bone and skin. Gelatin is natural protein produced from the partial hydrolysis of collagen present in bone and skin of animals [18]. The most common types of gelatine are manufactured from porcine skin, bovine bone, and bovine hide [19] (Fig. 6.4). Gelatines have been used as gelling agent, binder, emulsifier, or thickener. Pharmaceutical industries use gelatine for capsule cover, as drug deliver

Fig. 6.3 Products and by-products from livestock

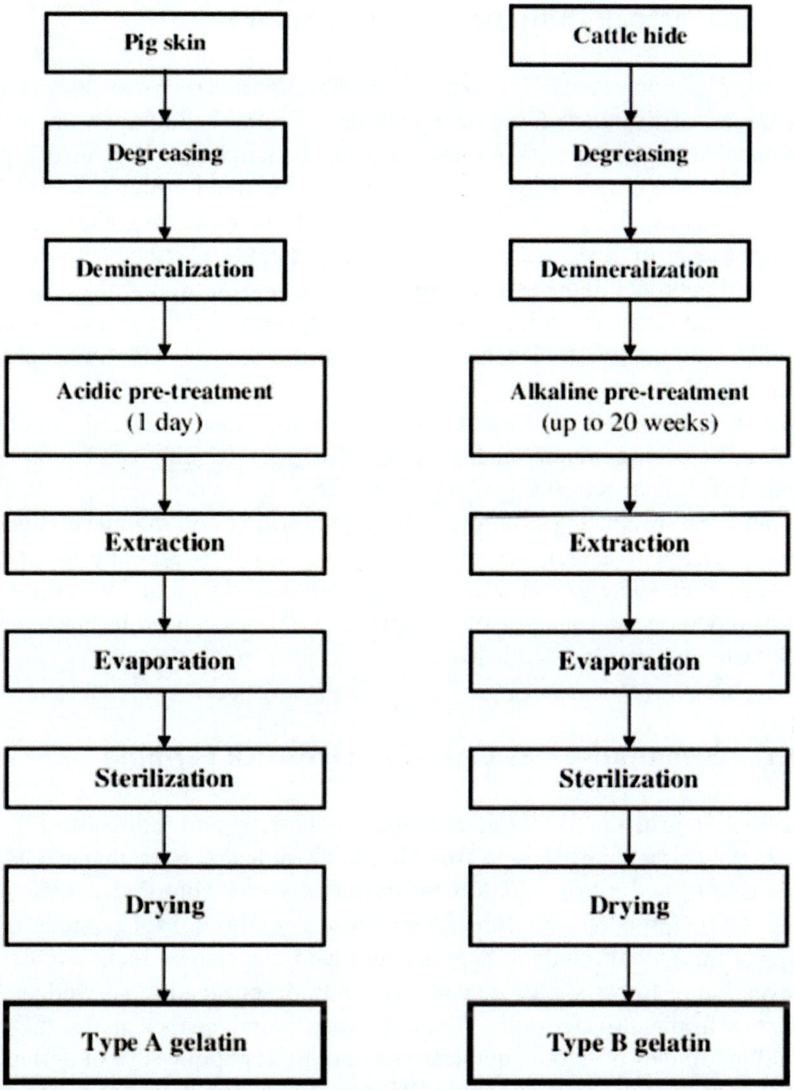

Fig. 6.4 Extraction of gelatine from pig skin and cattle hide

carrier. By the end of 2024, the global gelatine market will reach about US$3640 million. Gelatines are available in the market in variety of forms like Skin gelatine, Bone gelatine, and Hala gelatine [20].

6.3 Livestock Influencing Environment

Due to the growing demand for animal proteins, it is necessary to design and develop new model of livestock farmhouse in order to reduce the environmental problem associated with animal farming, and to bring sustainability in livestock production for better food security. The basic concept on integrated farming is to develop mixed faming system with concurrently activities involving crops and other animals, beside the cultivation of livestock for getting milk, meat, and other commodities. Such integrated approach reduces the burden of external resourcing like energy, fodders, and chemical fertilizer. In integrated farming system some crop used as animal feed or fodder, and the livestock generates bio-fertilizer (manure) for the crop growth and development.

But, with the development of modern technology the earlier concept of integrated farming has been gradually changing. The livestock farming can be integrated with biopharmaceutical industries to produce high value-added biologic drugs, besides the traditional products (milk, meat, fibre), and other commodities. However, mismanagement of livestock rearing contributes enormous pressure on environment, causes problems like global warming, land degradation, air and water pollution, and loss of biodiversity. Some of the important problems caused by mismanaged livestock farming are discussed below.

6.3.1 Greenhouse Gas Linked to Livestock Farming

Livestock industry produces greenhouse gas directly and indirectly (Fig. 6.5). The emission of greenhouse gases from livestock industry is methane (CH_4), nitrous oxide (N_2O), and nitrogen. Globally livestock sector contributes 18% (7.1 billion tonnes CO_2 equivalent) of global greenhouse gas emissions. Livestock is the major driver of the global trends in land use and land-use change including deforestation (conversion of forest to pasture and cropland), desertification, as well as the release of carbon from cultivated soils. CH_4 emissions from livestock are estimated at about 2.2 billion tonnes of CO_2 equivalent, accounting for about 80% of agricultural CH_4 and 35% of the total anthropogenic CH_4 emissions. The greenhouse gas emission to atmosphere is due to unusual management of livestock farming. Besides this, the greenhouse gas also contributes to land water degradation, biodiversity loss, acid rain, coral reef degradation, and deforestation. Ruminants have lower efficiency of N utilization and higher excretion of N in faeces and urine per unit N intake compared with non-ruminants [21].

FAO developed the concept of Livestock's Long Shadow (LLS) which is a life cycle assessment (LCA) of livestock's global impact on anthropogenic GHG emissions, biodiversity, land use, water depletion, water pollution, and air pollution [22]. LCA is a "compilation and evaluation of inputs, outputs, and the potential environmental impacts of a product or service throughout its life cycle" [23]. A LCA is a methodology used to assess both the direct and indirect environmental impact of a product from "cradle to grave". Environmental impacts that can be measured

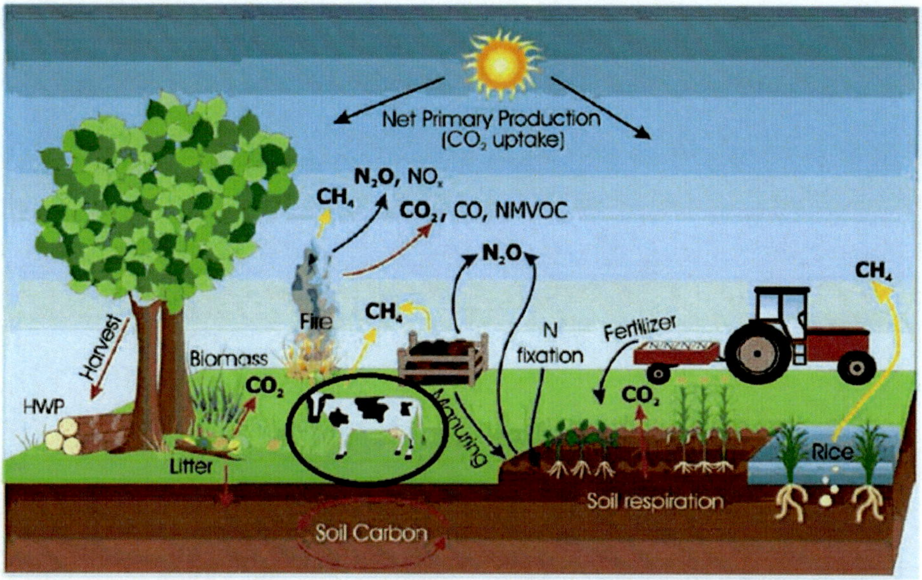

Fig. 6.5 Greenhouse emission linked to livestock (encircled in black)

include fossil fuel depletion, water use, GWP, ozone depletion, and pollutant production. The greenhouse gas emission in livestock production comes from enteric fermentation (as part of the natural digestive process of animals), manure management, and feed production for livestock. Direct emissions from livestock refer to emissions directly produced from the animal including enteric fermentation and manure and urine excretion [24].

Enteric Fermentation

Enteric fermentation occurs due to breakdown of biomass in rumen, due to action of bacteria, protozoa, and fungi. Ruminants convert cellulose and semi-cellulose into energy need of individuals (Fig. 6.6). Plant biomass in the rumen is converted into volatile fatty acids, which pass the rumen wall and go to liver through circulatory system. The enteric fermentation produces about 2.8 Gigatonnes which is about 39% of the total emission. A significant share of ruminants' environmental footprint is caused by enteric CH_4 that represents about >25% of the annual anthropogenic (manmade sources) CH4 emitted into the atmosphere with global dairy sector contributing 2.7–4% of the total anthropogenic GHG emissions [25].

Manure Storage

Manure acts as an emission source for both methane and nitrous oxide (Fig. 6.7). The emission depends on type of management and composition of the manure. Manure storage produced about 0.71 Gigatonnes and accounts for about 10% of total emission. Methane and nitrous oxide production in manure depends on organic matter and nitrogen content of excreta. Under anaerobic conditions, the organic

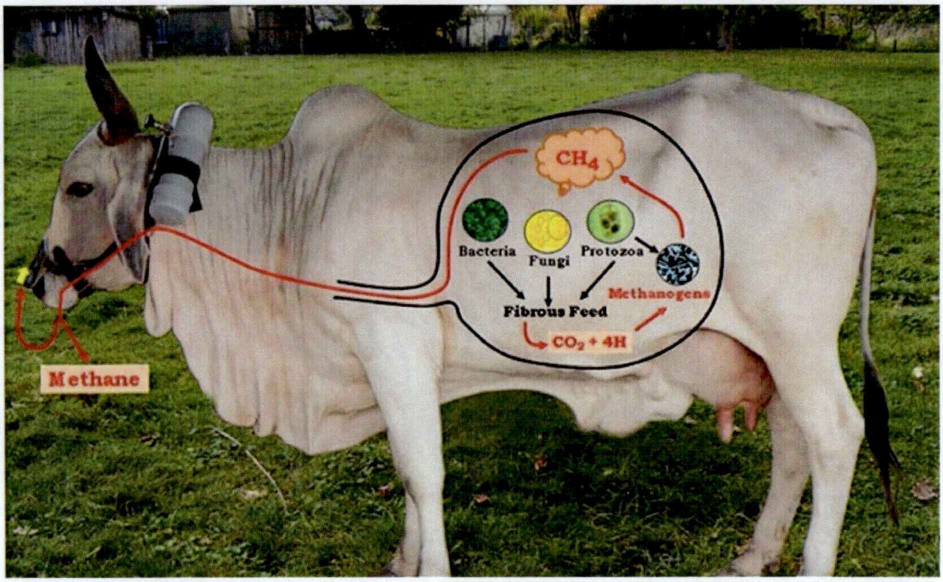

Fig. 6.6 Enteric fermentation and emission of methane gas causing greenhouse effect

matter is partially decomposed by bacteria producing methane and carbon dioxide. When manure is stored as a solid (dung) or deposited in pastures, nitrous oxide is generated through both the nitrification and denitrification process of the nitrogen contained in manure. Indirect N_2O emissions from livestock production include emissions from fertilizer used for feed production, emissions from leguminous feed crops, and emissions from aquatic sources following fertilizer application. The livestock sector contributes about 75% of the agricultural N_2O emissions (2.2 billion tonnes of CO_2 equivalent).

Feed Production

For livestock production, the term indirect emissions refer to emissions not directly derived from livestock but from feed crops used for animal feed, emissions from manure application, CO_2 emissions during production of fertilizer for feed production, and CO_2 emissions from processing and transportation of refrigerated livestock products [26, 27]. Other indirect emissions include net emissions from land linked to livestock including deforestation (i.e. conversion of forest to pasture and cropland for livestock purposes), desertification (i.e. degradation of aboveground vegetation from livestock grazing), and release of C from cultivated soils (i.e. loss of soil organic carbon (SOC) via tilling, natural processes) associated with livestock [26].

Feed production includes all the greenhouse gas emission arising from following activities such as:

- Land-use change
- Manufacturing and use of fertilizers and pesticides

Fig. 6.7 Greenhouse gas emissions from cow dung storage (common in Indian villages)

- Manure excreted and applied to fields
- Agricultural operations
- Feed processing
- Feed transport

Feed production accounts for about 45% of the total emission of whole livestock sector and is about 3.2 Gigatonnes of carbon dioxide equivalent.

6.3.2 Pastureland Sustainability

The demand for livestock products is increasing in both developed and developing countries due to increase in income level [25]. Economic factors are strong predictors of food consumption and the dietary intake of food components. In the developed countries, the consumption of animal protein is very high and more than 60% of protein comes from animal sources, whereas 20–25% of total protein comes from animal sources in developing countries including Africa. The demand will be exerted on limited resources and livestock rearing sustainability need to be analysed to fulfil the growing demand of animal protein in diet.

Pasture land is the farmland cover and used for animal production and comprising about 22–26% of earth's ice-free land surface [28, 29]. Livestock production can be increased by expanding the pasture land. But converting other land into pasture land, it will create different environmental problem. This needs to be analysed with broader perspective and also our preparedness to handle the problem. Similarly, it is required to increase grain production to feed the large growing population. Therefore, more land is required for grain production. The majority of global grasslands are located in the areas where it is impractical to cultivate the land for a variety of reasons. Grassland conversion to cultivated land will destroy the ecosystem, eliminate a major feed resource for grazing ungulates (including livestock), ruin the habitat for wildlife and other species, increase the risk of soil and wind erosion, increase nutrient runoff, and decrease soil carbon storage [30]. In short, the environmental risks are much too severe to convert a significant amount of grassland into cultivated cereals.

Pasture-based livestock system needs to be analyzed from social, economic, and environmental angle to access the sustainability of the system. Pasture land sustainability need to be analyzed at different spatial and temporal scale to understand the phenomena. Pasture or rangeland is commonly used for grazing animal. Pasture land refers to graminoids, herbs, and forages used for animal grazing, and land used permanently for herbaceous forage crops either by browsing or cultivating different forages. Browse land refers to trees, shrubs, and succulents that mainly used for browsing by livestock. Livestock mainly grazes for grasses, sedges, herbs, and forbs and browses for trees, shrubs, and succulents on browse land. Similarly foraging animals use lichens, mushrooms, nuts, and fruits in pasture land and browse land.

Pasture land improvement can be brought by integrating crop–livestock system. The integration of crop–livestock system can bring the stability and other benefits to pasture land. The crop based energy management model is proposed to highlight the benefit of crop-forage and livestock farming system. This system brings improvement of soil quality and provides more return to the farmers [31].

Pasture nutrient management can be another approach to bring sustainable improvement. Legume forages can be incorporated in grazing land, which provides better nutrition to livestock and improves the soil condition. Improved forage (legume) species in the pasture increases soil fertility by enhancing nutrient and organic matter levels [31]. Grazing management is another aspect to improve pasture, and careful analysis is required with respect to varied agro-ecological condition and soil type. The most important aspect of grazing management system is to optimize the grazing pressure with respect to land area [32]. Stocking density is another important parameter for pasture-based livestock production system.

Rotational grazing with proper stocking density improves soil quality, enhances root and shot growth, and decreases parasitic load.

Grazing by livestock has both positive and negative impacts on environmental quality management [33]. Generally, the impact of grazing of livestock is expressed on the basis of total protein production, nature of soil, and type of animal species grazing around green pasture. Grazing has dual effect on plant diversity.

Livestock consumes large quantity of feed which may result in negative impact on food security. The growing of cereals for feed in turn requires huge area of land. It has been estimated that a pound of beef (live weight) requires seven pound of feed, more than three pounds for a pound of pork, and less than two pounds per pound of chicken [34]. Feed quality is also responsible for production of live weight. For example, production of a pound of beef cattle live weight may require 4–5 pound of feed containing high protein [35].

Generally, post-weaning growth rate is less than the pre-weaning growth. Post-weaning ADG is a measure of individual animal's performance and rearing environment. Growth after weaning depends on energy intake by an individual. For example, the rate of tissue accumulation and growth rate both decline as maturity is approached. Production of commodities (meat, milk, etc.) is mainly dependent on the types of livestock.

Agriculture land around the world is limited and the processes of urbanization gradually grabbing the agriculture land and may create alarming condition in future. Due to the limitation of grazing land and increase in the demand of animal protein, continuous grazing has been a challenging issue on stability and diversity of green land ecosystem. Grazing can damage habitats, destroy native plants, hamper plant diversity, and cause for soil erosion. In addition, it has also been noticed that livestock eat selective plants which may cause the increase in biodiversity of wild plants and weeds those are responsible for bringing deficiency in soil nutrient value. In annual cropping system communities, free grazing livestock often damages crops and is a major cause of conflict. So, it has been realized that the practice of zero grazing is helpful in conserving the communal grazing land. In order to overcome with the problem of continuous grazing, the concept of zero grazing gives challenge for better nutritional management for more milk productivity. Zero gazing is a form of dairy farming where cattle are kept within housing structure; forage is arranged for them either by developing hydroponic culture (Fig. 6.8) of different cereals or grass.

6.3.3 Bioaccumulation of Antibiotic

Antibiotics are mainly used in livestock industry to prevent the diseases as well as growth promoter. The antibiotics are used for disease treatment, prevention, and growth promotion. The most commonly used antimicrobials in animal production are penicillins, macrolides, polypeptides, streptogramins, and tetracyclines. Other antibiotics such as quinolones, lincosamides, and aminoglycosides are primarily used only in disease treatment or prevention. Due to widespread use of antibiotics for treatment, prophylaxis, and growth promotion, livestock has become a reservoir

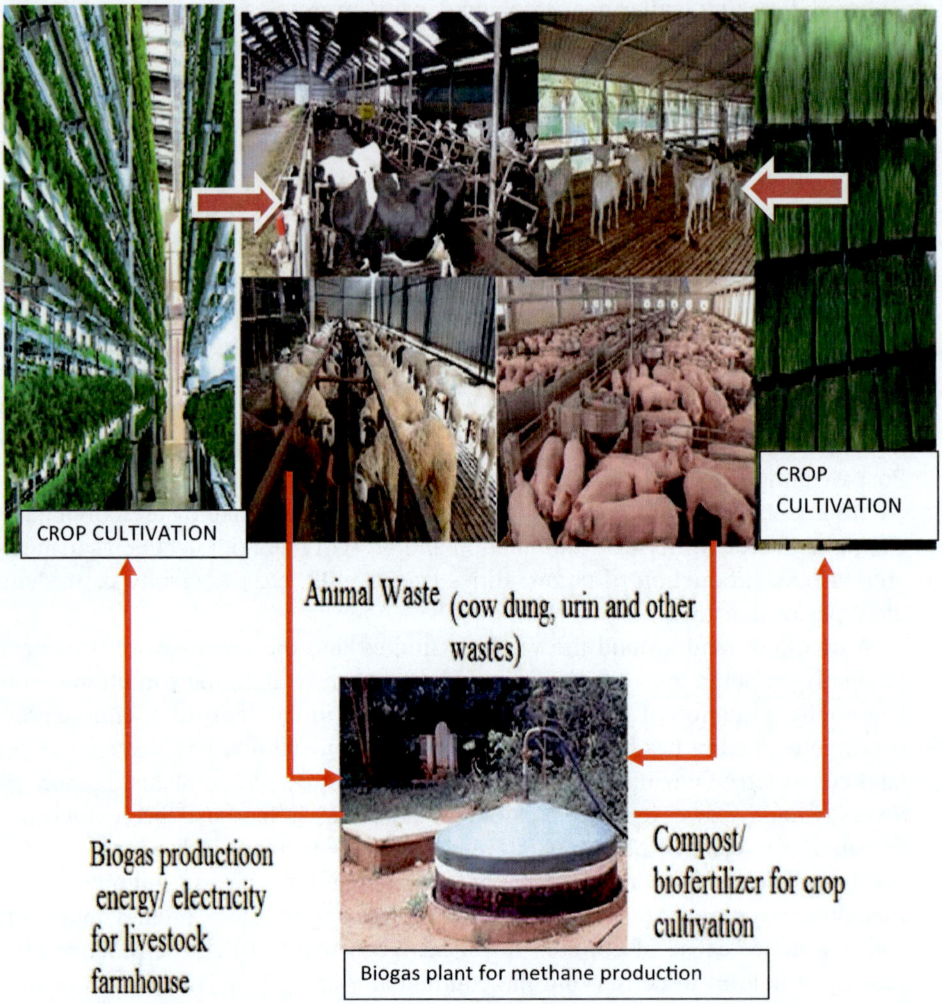

Fig. 6.8 Sustainable livestock farming with zero greenhouse gas emission and recycling of animal waste for in house farming

of antimicrobial-resistant bacterial strains and genes. The antimicrobial-resistant strains transmit directly to farmers and veterinarians and to consumers through food chain [36, 37]. The antibiotic residues in excrement may cause high level of antibiotic accumulation in the environment (Fig. 6.9). Antimicrobials are delivered to animals for a variety of reasons, including disease treatment, prevention, control, and growth promotion/feed efficiency. Antimicrobial growth promotants (AGPs) were first advocated in the mid-1950s, when it was discovered that small, sub-therapeutic quantities of antibiotics such as procaine penicillin and tetracycline (1/10 to 1/100 the amount of a therapeutic dose) delivered to animals in feed could enhance the feed-to-weight ratio for poultry, swine, and beef cattle [38]. For many years, the positive effects of this practice were championed, while the negative

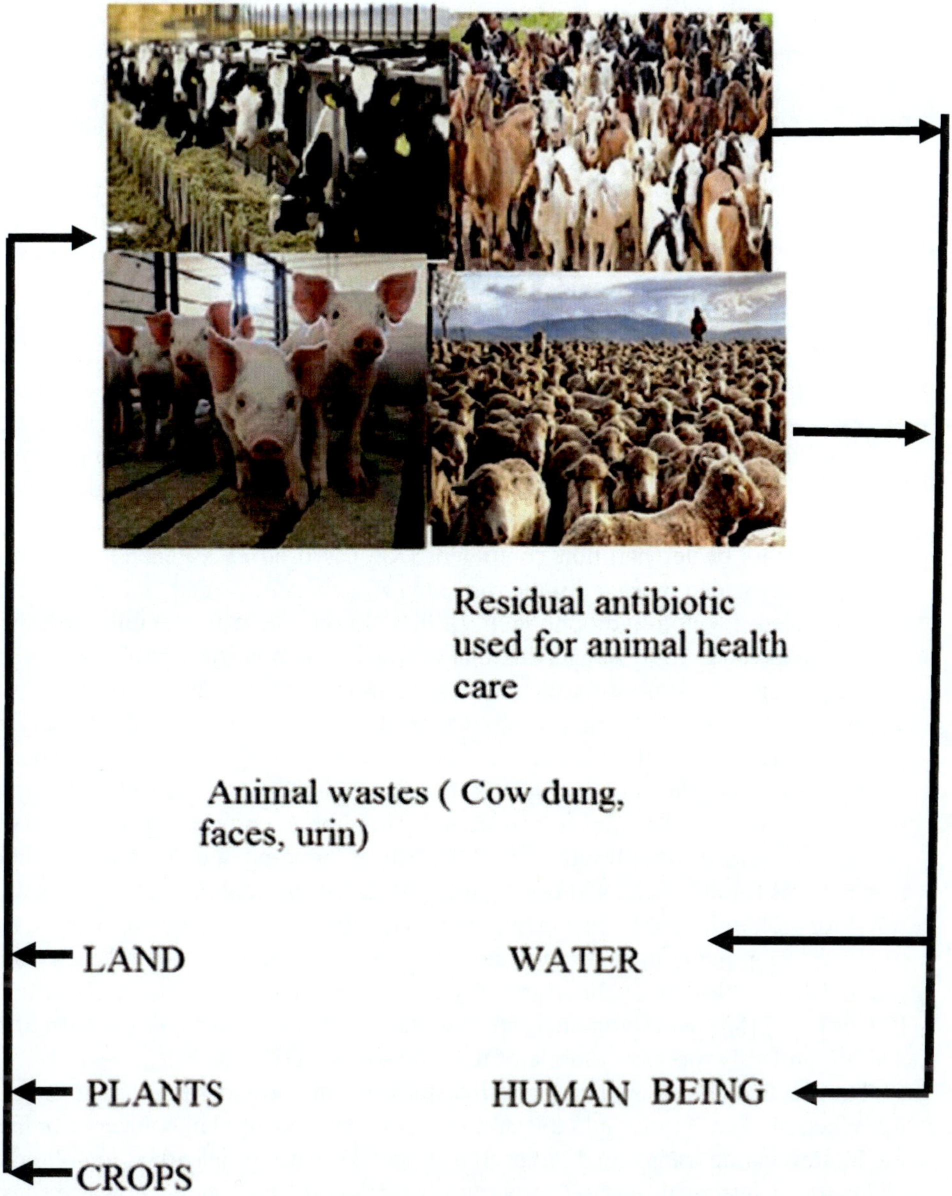

Fig. 6.9 Residual antibiotic (leaching from livestock farmhouse) contamination to human being

consequences went undetected. But microbiologists and infectious disease experts facing antibiotic resistance questioned the possible harm from this use [39–42]. They found that farms using AGPs had more resistant bacteria in the intestinal floras of the farm workers and farm animals than in those for similar people and animals on farms not using AGPs.

Many antibiotics are not completely absorbed in the gut, resulting in the excretion of the parent compound and its breakdown metabolites [43–45]. The antibiotics are excreted into environment in various ways. The antibiotics are excreted to environment through urine, faeces, atmospheric dispersal of feed, and manure dust containing antibiotics.

Biodegradation of organic compounds by microorganisms in soil is dependent in part on factors such as temperature, concentration, bioavailability, and time of exposure, availability of other nutrients, and the enzymatic capabilities of the extant microbial population. Aerobic processes have been the primary focus of such studies, and little attention has been devoted to anaerobic processes, the latter being of significance in the soil subsurface and microzones. Quinolones and tetracylines are susceptible to photodegradation [46], and photodegradation of oxytetracycline is three times more rapid under light than dark conditions [47]. Halling-Sørenson [48] suggested that tylosin might be resistant to photolysis because it has only limited light absorbance in the visible spectrum, and Boxall et al. [45] determined that sulphonamides would not be readily photodegraded.

The octanol–water partition coefficient (K_{ow}) is used as a general measure of hydrophobicity, and most antibiotics have log K_{ow} values less than 5, indicating that they are relatively non-hydrophobic [49]. In addition, the water solubility for many antibiotics exceeds 1 g/L, suggesting that they are relatively hydrophilic. Antibiotics exhibit a range of affinities for the solid phase (K_d 0.2–6000 L/kg), with consequent effects on their mobility in the environment. Estimations of antibiotic organic carbon-normalized sorption coefficients (K_{oc}) made by using a compound's octanol–water partition coefficient (K_{ow}) generally result in underestimates of the K_{oc} value, suggesting that mechanisms other than hydrophobic partitioning occur. Cation exchange, surface complexation, and hydrogen bonding are included as likely mechanisms for antibiotic sorption to soils. Many of the acid dissociation constants (pK_a) for antibiotics are in the range of soil pH values, such that the protonation state of these compounds depends on the pH of the soil solution [49]. Tetracyclines (tetracycline, chlortetracycline, and oxytetracycline) can adsorb strongly to clays [50–54], soil [55], and sediments [56]. Sulphonamides exhibit weak sorption to soil and are probably the most mobile of the antibiotics [49].

The practice of land application of livestock manure provides large-scale areas for introduction of antibiotics into the environment. Once released into the environment, antibiotics can be transported either in a dissolved phase or adsorbed to colloids or soil particles into surface water and groundwater [57–60]. Manure and waste slurries potentially contain significant amounts of antibiotics and their presence can persist in soil after land application [61, 62]. Antibiotics excreted from animals are often concentrated in the solid phase because of adsorption dynamics [49, 63–65]. Gavalchin and Katz [62] determined the persistence of seven antibiotics in a soil-faeces matrix under laboratory conditions and found that the order of persistence was chlortetracycline > bacitracin > erythromycin > streptomycin > bambermycin > tylosin > penicillin with regard to their detection in the soil. The application of manure to agricultural fields also likely introduces breakdown products into the environment along with the parent compound, but persistence data for degradation products were not found in the literature reviewed.

The livestock industry is showing increasing antimicrobial resistance in similar manner happening to human clinical isolates [66]. Antibiotics are not degraded in the animal body and thereby enter into environment through excreted urine and faeces [67]. The livestock farms have been reported to discharge a large amount of antibiotics into nearby surface water and soil [68, 69]. Residual antibiotics usually have negative impacts on organisms, food security, and water quality, which in turn accumulate in human body through the food chain and drinking water [70]. The problem is evident in both developed and developing countries. The maximum concentration of oxytetracycline was 340 ng/L in the United States [58] and similarly the maximum concentration of sulfadiazine and oxytetracycline was 4130 ng/L and 4490 ng/L, respectively, in the UK Environmental Agency. The emergence and spread of agricultural antibiotics are matter of concern and farming practices are largely are to be blamed for the rise of antibiotic-resistant strains [71]. A human is infected by a resistant pathogen of agricultural origin through contact with livestock, or through ingestion of bacteria from contaminated meat or water, without ongoing transmission of the pathogen between humans. Resistance genes arising in the agricultural setting are introduced into human pathogens by horizontal gene transfer. The resulting resistant lineages are then selected by antibiotic use in humans.

Animal husbandry is a major food producing industry worldwide. Poor gut health is associated with infection, inflammation, intestinal atrophy leading to stunted growth, morbidity, and mortality in livestock. A prospective in vivo/in situ study in 1975 was performed to evaluate the effect of introducing low-dose in-feed oxytetracycline as an AGP on the intestinal floras of chickens and farm dwellers [72]. The results showed not only colonization of the chickens with tetracycline-resistant and other drug-resistant *Escherichia coli* strains but also acquisition of resistance in *E. coli* in the intestinal flora of the farm family. Other studies over the ensuing three decades further elucidated the quantitative and qualitative relationships between the practice of in-feed antimicrobials for animals and the mounting problem of hard-to-treat, drug-resistant bacterial infections in humans. Antimicrobial resistance poses a worldwide threat to public health, which may be partially associated with using AGP in livestock production. The European Union banned the use of AGP in animal food production in 2006. The US FDA placed restrictions on antibiotic use in animal production in December 2016. Antibiotic resistance among commensal bacteria represents a major avenue for the development of resistance in bacterial pathogens, since resistances increase first in commensals and are then transferred to pathogens later. Commensal gut bacteria are likely to be highly efficient contributors to antibiotic resistance because the numbers of commensal bacteria in the intestinal ecosystem are large, often more than 1014 bacteria comprising several hundred species [73].

6.4 Measures to Prevent Adverse Environmental Impacts

Generally, livestock production is adversely affected by various environmental problems. It is required to prevent the damage of ecosystem surrounding the livestock farming which ultimately affects the animal protein productivity and food security. The appropriate measures should be taken to improve the sustainability of livestock production system. Climatic changes accompanied by population growth, environmental degradation, poverty reduction, and food insecurity are challenging task during the century. The first agreement on climate change was adapted at the UN climatic Change Conference of Parties in Paris in 2015. About 195 countries participated to bring global awareness about the alarming climatic changes caused by environmental problems, with special reference to greenhouse emission. It has been observed that dairy sector is one of the major contributors for greenhouse gas (GHG). Since the last decade attempts are in pathway for monitoring GHG emission from livestock farmhouses by introduction of integrated farming technology, zero grazing process, and recycling of livestock wastes for generating biogas and bio-fertilizer. The emission intensities of GHG per kilogram of milk have declined by almost 11% over the period 2005–2015. Generally, livestock contributes two-thirds of agricultural GHG emission and 78% methane emissions. On the basis of Global livestock Environmental Assessment Model (GLEAM) the emission from feed crops account for about 45% of total emission and methane and nitrous oxide from manure reach about 10%. As per FAO assessment, this emission is about 30% of baseline emission.

It has been identified that Sub-Saharan Africa and South Asia are the most vulnerable regions for climatic change. In these region farmers and rural communities rely mostly on livestock for food, income, and livelihood. Natural calamities like flood, extreme drought hamper the productivity and are also responsible for the quality production of forages. FAO helps in monitoring the livestock production under extreme climatic change. In addition, FAO also helps in identification of interventions to increase productivity and reduce the impact of climate variability on livestock outputs, including adaptation needs. FAO attempts to develop information databank in public domain to explain updated information on development of tools for practice change and engages in policy dialogues. In such activities, the Global Agenda for Sustainable Livestock acts as multi-stakeholder partnership with FAO to bring together private sector, governments, civil society, research and international organizations to a common platform. Livestock Environment Assessment and Performance (LEAP) with FAO partnership has developed common metrics and methodologies to explain and measure environmental performance and GHG emission in livestock supply chain, and reduce enteric methane for improving food security and livelihoods project with the Climate and Clean Air Coalition.

6.4.1 Mitigation of Greenhouse Gas from Livestock

Information on mitigation practices for greenhouse gas emissions (non-carbon dioxide) from livestock is scanty. However, consolidated information on existing greenhouse gas mitigation practices applicable to different livestock systems across the globe is at initial stage of development.

On the basis of data available from Intergovernmental Panel on Climate Change (IPCC), Food and Agriculture Organization (FAO), Environmental Protection Agency (EPA) or other agencies livestock contribute to global anthropogenic GHG emissions and varies from 7% to 18%. It is mainly due to uncontrolled nutrient input and animal husbandry practices. In order to mitigate the methane (CH_4) and nitrogen oxide (NO_2), the non-carbon dioxide emission emphasis has been given on enteric CH_4 mitigation practices for ruminant animals and manure mitigation practices for both ruminant and non-gastric species.

The digestible forage is given to livestock to reduce the GHG emissions from rumen fermentation. This practice is highly recommended as mitigation practices. Enteric CH_4 emission may be reduced by replacing corn silage in place of grass silages in the diet. Legume silages are also a better option in place of grass silage due to their lower fibre content. Grass pastures with legumes, in warm climate condition offer a mitigation opportunity. Dietary lipids are also reducing the enteric CH_4 emissions. It has also been noticed that inclusion of concentrate feeds in the diet of ruminants will likely decrease enteric CH_4 emission. Nitrates are also proven to be effective enteric CH_4 mitigation agents, particularly in low-protein diets that can benefit from nitrogen supplementation. But, adaptation to these compounds may be toxic. Tannin may also helpful in reducing enteric CH_4 emissions, but it may interfere in overall milk production. Some direct-fed microbes, such as yeast-based products help in moderate CH_4 mitigation effect by increasing animal productivity and feed efficiency.

6.4.2 Mitigation Practice Through Manure Monitoring

Dietary composition has impact on manure quality production (faeces and urine). So, monitoring of GHG emission during the storage of manure is also an important factor for climatic change. Generally, manure storage is required where animals are housed indoors or on feedlots. But higher proportion of ruminants is grazed on pastures or rangeland, where CH_4 emission from their excreta is very low and N_2O losses from urine can be substantial. Sometimes, inefficacy in digestion leads to increase in fermentable organic matter concentration in manure which may be ultimately responsible for manure CH_4 emission. It is necessary to have dietary balance in low-protein intake for ruminant so that microbial protein synthesis and fibre degradability are not impaired. Reducing in total dietary protein and supplementing the diet with synthetic amino acids are effective on ammonia and N_2O mitigation for non-ruminants. It has been understood that overall process on housing, type of manure collection and storage system, separation of solids and liquid and their processing have remarkable impact on ammonia and GHG emission. Mostly, mitigation for GHG can be controlled by reducing the time of manure storage, aeration, and stacking. It is hard to assess economic aspects of these types of mitigation. Certain physical practices like covering the manure with impermeable membranes, such as oil layers and sealed plastic covers, are effective in reducing gaseous emission. Accumulation of CH_4 and subsequent utilization for electricity production is a better option to meet mitigation of GHG from livestock manure. Acidification (in areas where soil acidity is not an issue) and cooling are further effective methods for reducing ammonia and CH_4 emissions from stored manure. Generally, anaerobic digestion of cattle manure is recommended for CH_4 generation as alternate energy and mitigation of GHG emissions. However, anaerobic digestion is not suitable for the geographical location below 15 °C.

6.4.3 Animal Productivity and Mitigation Practices

Animal productivity is also a vital factor for monitoring GHG emission per unit of livestock product. The mitigation of GHG emissions can be approached by genetic engineering and breeding practices. Control over herd size is helpful in the availability of adequate feed to livestock at individual level which will ultimately helpful in lowering CH_4 emission. Reducing the age for slaughtering with controlling dietary plan also has a significant impact on GHG emission.

Forage quality, dietary nutrient controlling practices are effective ways of reducing GHG emission in livestock farmhouse. Varieties of feed supplements have potential impact on reducing GHG emission per unit animal product.

6.4.4 Social Dimension

It is true that rural economies, especially of developing nations, also depend on animal production through livestock farming [74]. In some geographic location like Mediterranean basin, livestock production system is only the alternate way of managing livelihood [75, 76]. From cultural hierarchies prospective, the specific choice of livestock is crucial with high aesthetic and environmental value [77, 78]. Organic livestock farming is having lot of environmental benefits, which could save lot of budgetary provision for reducing greenhouse effect, conserving plant diversity, and maintain nutrient values of soil [79, 80].

Livestock supports the livelihood of poor farmers, consumers, traders, and labourers throughout the developing world. It is required to enhance livestock production system for a better output. Livestock business is a flexible practice in which the rural people can convert it into cash on family need base, either by directly selling the animal or on informal credits and loan basis. In rural locality, where provision of financial market is not available, livestock or herds can be used as source of collateral security for getting loan. Livestock is a wonderful saving account for rural people whenever they want to meet their financial deficit budget under

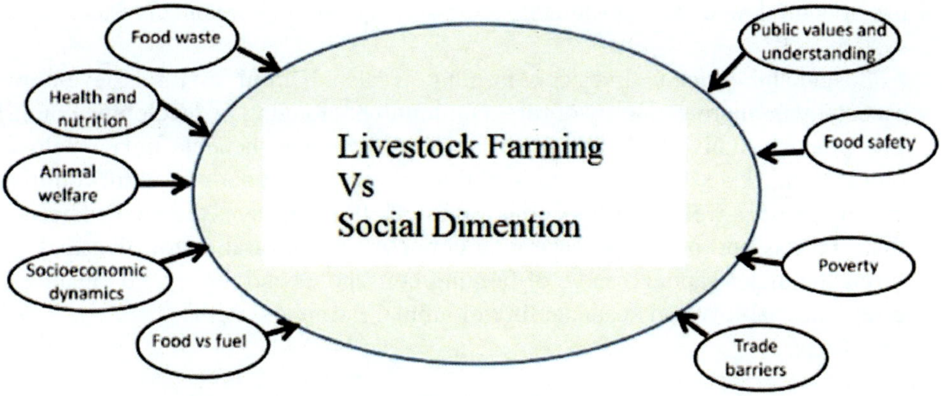

Fig. 6.10 Livestock farming associated with social dimension

unavoidable circumstances like low crop production or failure due to change in climatic condition. So, livestock is a valuable source of risk reduction and security for rural people. Organic livestock farming is one of the best sources for chemical free fertilizer to be used as manure for organic cereal or vegetable or fruits production, and having high price as compared to conventional pattern of food crops production. Organic livestock practice is not only eco-friendly, but also improve energy and nutrient cycling, and can also be used to transport agricultural inputs and output. In addition, livestock plays significant role in boosting and strengthening rural market. Rural markets are an integral part of rural community financial flow and well-being, wealth, and as a result provide social status to livestock owners by an additional budget to meet the expenses for fulfilment of a set of rituals, and social obligations of families and communities.

Therefore improvement of rural area has become a challenge for developing nations, in order to bring socio-economic stability, quality of life, and sustainable in agricultural productivity (Fig. 6.10). But, the concept of sustainable rural development is not yet clear. The basic target of sustainable development means process of change and multidimensional evolution related to social inclusion, rural environment conservation, rural micro economic, and prevailing political situation. In this context, integrated organic livestock farming has high potential to contribute to the development of rural areas. But, over the years, the rural farmers have been practicing the rational type of agriculture practice and animal husbandry due to lack of information and training on integrated organic livestock farming or multi crops organic farming. Acquisition of such primitive skill by our rural farmers has not helped to improve agricultural yield or livestock production.

Livestock extension education in rural village communities is an important aspect for pursuing the farmer to practice integrated livestock farming. In this connection the farmers and rural women are to be acquainted with the use of local renewable resources (windmill energy, solar energy, and biogas production) as sources of energy, primarily necessary for organic livestock farm development. Sustainable use of natural renewable resources is crucial in developing organic livestock farming system.

Sustainable organic livestock farming is also helpful in rural development, especially in micro-economy of rural community belonging to developing countries. The practice of livestock farming is having diversified benefits linked with food production, food security, provides additional economic goods for rural community development as well as cash income. Livestock farming saves the rural society from financial risk and provides year-round employment. In rural sector, livestock often forms the major capital reserve of farming households and, in general enhances the economic viability and sustainability of animal farming system.

6.4.5 Sustainable Organic Livestock Farming

Sustainable livestock farming is one of the best ways to challenge the future food security. Organic livestock production is targeted to produce variety of food, besides animal proteins with some stipulated policies associated with better animal welfare, care for the surrounding environment, minimum use of chemical fertilizers, restricted use of drugs, and production of healthy products without pesticide residue or any harmful chemicals. Commonly, in organic livestock farming, the animal's feedstock is cultivated in the field in which feedstock is grown with the input of manure obtained from animals. It is strictly advised not to use any chemical fertilizer or pesticides. The application of chemical fertilizer can increase the crop yields quickly, but promote in soil hardening and reduce the content of soil organic matter and in long run loss spoil productivity potential. Livestock manure not only contains organic matter but is adequately rich in essential elements, and micro- and macronutrients. The factors associated with organic farming are described in Fig. 6.11.

6.4.6 Generation of Energy and Organic Manure

Livestock farm (dairy farm) can be independent from conventional energy (Electricity) to meet overall energy expenditure for providing electricity throughout the year. This could be possible by linking livestock farmhouse waste with biogas production. Even other organic wastes generate from farmhouse can be used for biogas production along the cow dung.

Biogas generated from animal waste can be used in similar way as natural gas for cooking, running freeze, fans, and lighting bulbs. Raw biogas is composed of 50–75% methane, 25–45% carbon dioxide, 2–8% water vapour, and traces of O_2, N_2, NH_3, H_2, and H_2S. All the unwanted gas from biogas can be removed by water scrubbing, pressure swing, adsorption (PSA), amine gas treatment, and CO_2 cooling and recovering dry ice. Among these methods, PSA technology is commonly used for removing CO_2 gas. The average calorific value of biogas is about 21–23 MJ/ma, so that $1m^a$ of biogas corresponds to 0.5–0.61 l diesel fuel or about 6 kWh. The overall productions of biogas also depend on the nature of feedstock and design of

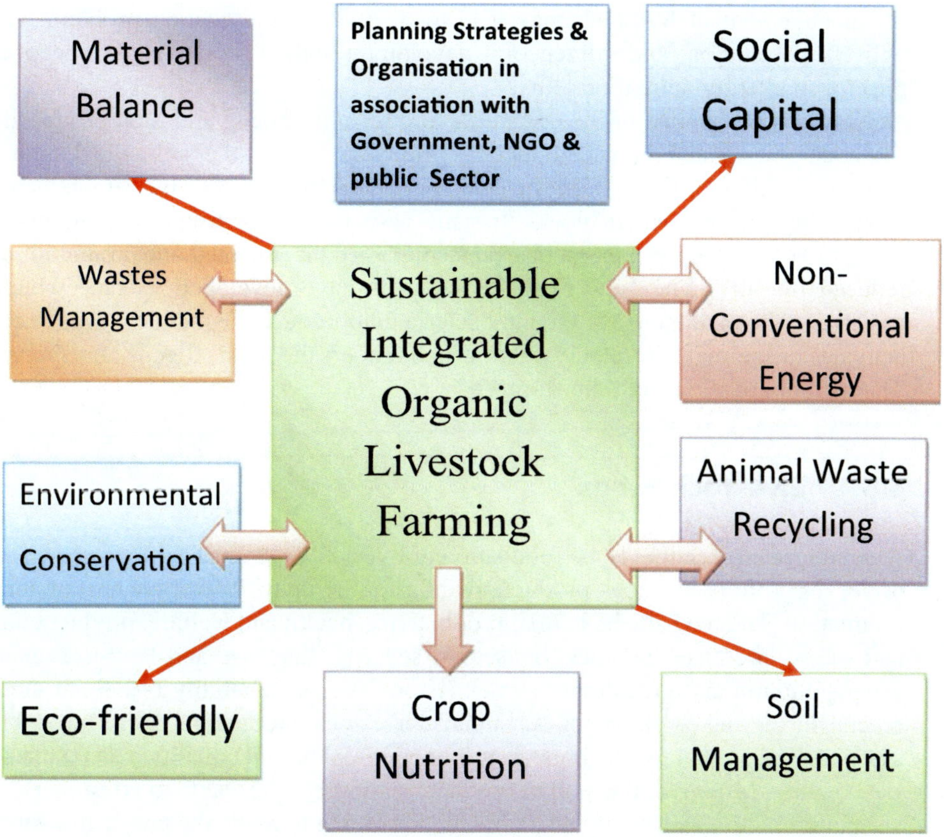

Fig. 6.11 Factor associated with organic livestock farming

the biogas plant. Maize silage, for example, (a common feedstock in German) yields about 8 times more biogas per ton than cow manure. The biogas production also depends on the type of animal from which the waste is collected. Following are the tentative biogas production yield from different animal wastes:

- 1 kg cattle dung 40 L biogas
- 1 kg buffalo dung 40 L biogas
- 1 kg pig dung 40 L biogas
- 1 kg chicken droppings 70 L biogas

6.5 Good Animal Health

Maintaining good animal health is the foremost requirement in developing a sustainable pattern of livestock farming. Good animal husbandry practices minimize impact on the environment. So, in 2013, FAO announced that by adapting existing best practices in health and husbandry, the GHS emission can be reduced by 30%.

Mismanagement of livestock may lead to diseases contamination problem which will effect lactation, and influence in developing animal (cows, goats, sheep, and pig) for marketing animal proteins.

Nutrition is also one of the critical factors for minimizing emission produced by livestock. Good and balanced dietary provision boosts the animal's health. In addition, methane production in livestock can be reduced by supplementing 3-NOP in the feed which blocks enzyme associated with triggering microbes in the gut to create the methane. It has been confirmed that the methane production was reduced by 30%. The New Zealand scientists have developed vaccine which is targeted to the methanogens, the gut bacteria that produce the methane. The vaccine activates the animal's immune system which renders the methanogens unresponsive. They simply pass through the stomach.

6.6 Livestock Feeds

Animal feed and feeding is the foundation of livestock systems. The feed is the major driver for profit in livestock production and forming up to 70% of the cost of animal production. The feed affects animal productivity, health and welfare, product quality and safety, producer incomes, household security, land use and land-use change, water pollution and greenhouse gas emission. The sustainability of animal diets is crucial in the development of sustainable livestock production across production systems. The animal feed is important as the animal product quality is dependent on feed quality. Moreover the animal products should be affordable to consumers. It is very important to use locally available feed resource to make the production system more reliable and profitable. The challenge is to identify novel, alternate feedstuffs that do not compete with human food and their incorporation in the diets results in feed use efficiency as high or greater than that obtained using grain-containing animal diets currently in use. It has been showed that decreasing grains and increasing fibre in the diets of animals including monogastrics enhance both animal welfare and animal production [81–83]. The livestock production is highly variable with respect to food regime and regions. Pasture-based system has potentially beneficial role as they contribute to balanced nutrient recycling ecosystem [84]. Similarly legume based pastures are biological source of nitrogen for soils and protein for livestock [85].

The profit dimension is very important; however, environmental aspect should be considered for livestock production. The profit is important since it is the main driver of the production system, but making profit at the cost of the environment, socio-economic benefits to people and animal welfare, and other ethical aspects of raising livestock may not be appropriate. Decrease in importance to profit dimension does not imply that livestock farming will be less profitable. The profitability could be achieved by better consideration of the other dimensions of the sustainable development.

According to the Environmental Working Group (EWG), numerous studies indicate grass-fed beef has less saturated and mono-unsaturated fat and more nutrients than grain-fed beef. Consuming organic, grass-fed, and pastured-raised livestock also reduces eaters' exposure to antibiotics, pesticides, and hormones, which are used in conventionally raised livestock.

6.7 Livestock and Global Food Security

The demand for food is expected to double by 2050. World agricultural systems will have to face the challenges to feed the growing population. The population growth will provide opportunities and challenge for the dairy sector. Global Dairy Market Report and Forecast says that the global dairy market is supposed to grow by 2.5% per annum till 2020. This is mainly due to the nutrient value of dairy products rich in proteins, vitamins (riboflavin, vitamins B5 and B12), and minerals like calcium, magnesium, selenium. Dairy products are the fifth largest provider of energy and third largest provider of proteins and fat for human beings. It has been surveyed that more than 150 million farmers possess at least one milk cattle (Cow, buffalo, goat, or sheep). Worldwide, 133 million holdings keeping dairy cattle, 28.5 million with buffaloes, and 41 and 19 million with goats and sheep, respectively [86]. Generally, farmers prefer to keep mixed herds with more than one species of dairy animal. Cows are noticed to be the most common in farmers of developing countries [86]. It is interesting to note that about 25% of cattle-keeping householders, or in about 35 million farms, dairy cows are directly owned and/or managed by women [86]. Mostly, in rural area women of developing countries feel safe and secure by having themselves by possessing cattle and some other livestock helpful in farmland management. Still, rural women community faces challenge in marketing dairy products at competitive prices when dairy production is subjected to changing weather patterns, changing market dynamic and prices. Livestock are important contributors to total food production both for developed and developing countries. The rate of livestock production is at higher magnitude as compared to cereals. Livestock are primary source of proteins and essential amino acids, and major contributor for human diet as source of energy.

6.8 Association of International Organizations with Livestock Monitoring

Livestock provides nutritious food, fuel, transportation, and other products. Livestock is associated with livelihood security of farmers in developing countries. Therefore, there are many international rural organization involved in up-gradation of livestock farming system for better sustainable healthy life.

The Intergovernmental Panel on Climate Change

The Intergovernmental Panel on Climate Change (IPCC) is an intergovernmental body of the United Nations dedicated to provide information and help for management of livestock farming in a sustainable form. It also helps in understanding the risk of human-induced climate change, its natural, political and economic impacts and risks, and possible response options. The IPCC was established in 1988 by the World Meteorological Organization (WMO) and later authorized by United Nations Framework Convention on Climate Change (UNFCCC), the main international treaty on climate Change. The main target of this international organization is to monitor and bring stability in GHG emission to atmosphere at a level that would prevent dangerous anthropogenic interference with climate system. The IPCC is an internationally recognized body on climate change involved in publishing report on public domain.

The Food Agriculture Organization of the United States

The Food Agriculture Organization of the United States (FAO) is an international agency involved to eradicate hunger from global map by improving nutrition quality and food security. Its Latin motto, *fiat panis* stands for "let there be bread". The FAO's headquarter is located in Rome, Italy and maintains regional and field offices around the world, operating in over 130 countries. The main target is to coordinate the activities to upgrade agricultural practices, livestock management, forestry, fisheries, and land and water resources. In addition, it provides technical assistance to work out projects and to promote food security.

The main role of FAO is to identify and asses the livestock condition, and to advice for necessary steps to be taken for livestock management and sustainability. The Global Agenda for Sustainable Livestock, with the help of FAO generate information on for practice policy dialogue with policy makers and government on livestock management. In addition it gives special emphasis on social, economic, and environmental performance of livestock systems.

Earth University (Costa Rica)

This university is actively involved in developing sustainable animal husbandry practices, providing training to farmers and other visitors on sustainable livestock management, including the recycling of animal wastes through anaerobic fermentation to generate energy in the form of methane gas.

The Human Society of the United States (HSUS)

This society provides guidelines on livestock development guidelines, and also campaigns reform of existing traditional livestock farmhouses as sustainable pattern. HSUS and their associate partners provide care to more than 100,000 animals per year.

The International Livestock Research Institute (ILRI)

ILRI works to improve food and nutritional security and reduce poverty in developing countries through research for efficient, safe, and sustainable use of livestock. It is the only one of 15 CGIAR Research Centres dedicated entirely to animal agriculture research for the developing world. Co-hosted by Kenya and Ethiopia, it has regional or country offices and projects in East, South, and Southeast Asia as well as Central, East, Southern and West Africa. www.ilri.org. The ILRI also helps the poor farmers to use livestock as tool to secure assets and develop marketing production channel to improve economic status of smallholders and pastoral community.

The Natural Resources Defense Council (NRDC)
The NRDC helps in taking action to prevent superbugs in meat production. In addition they are involved in promoting healthy chickens, pigs, and cows to improve health and minimize environmental damage causes due to mismanaged livestock farming system.

Joint FAO/WHO Meeting on Pesticide Residues (JMPR) and The Joint Meeting on Pesticide Specifications (JMPS)
JMPR is an international expert scientific group administered jointly by the Food and Agriculture Organization of the United Nations (FAO) and WHO. JMPR meets regularly since 1963 to review residues and analytical aspects of the pesticides, estimate the maximum residue levels, review toxicological data, and estimate acceptable daily intakes (ADIs) for humans of the pesticides under consideration. This initiative is for reviewing residue and analytical aspects of the pesticides, including data on their metabolism, fate in the environment and use patterns, and for estimating the maximum levels of residues that might occur as a result of use of the pesticides according to good agricultural practice.

The Joint Meeting on Pesticide Specifications (JMPS) is an expert ad hoc body administered jointly by FAO and WHO, composed of scientists collectively possessing expert knowledge of the development of specifications. Their opinions and recommendations to FAO/WHO are provided in their individual expert capacities, not as representatives of their countries or organizations. The primary function of the JMPS is to produce recommendations to FAO and/or WHO on the adoption, extension, modification, or withdrawal of specifications and to develop guidance and procedures in establishing pesticide specifications and equivalence determination which has also its relevance to the registration and quality control of pesticide in national or regional authorities.

6.9 Sustainable Development Goals and Livestock Management

The sustainable development goals (SDGs) are collection of 17 goals designed to be a "blueprint to achieve a better and more sustainable future for all". The SDGs set in 2015 by the United Nations General Assembly. The main target includes 17 key elements to improve health, education, the environments, gender equality, and food security, by the year 2030 Livestock species conservation, monitoring, and faming are the few important aspects of SDGs programme to be targeted by the end of 2030. The Livestock plays a vital role in economic, social, and cultural function for rural households in developing countries. Livestock has multifunctional impact on the development of food supply, family nutrition, family income, asset saving, soil productivity, livelihoods, transport, agricultural traction and diversification. The

SDGs efforts are to develop economic infrastructure at rural level, increase and strengthen rural supply chain management. The SDG effort is to strengthen livestock supply chain and contribute significantly to family nutrition by supplying animal protein. The SDGs strategies are to increase household consumption of protein principally from animal origin.

References

1. Godfray HCJ, Beddington JR, Crute IR et al (2010) Food security: the challenge of feeding 9 billion people. Science 327:812–818
2. Food and Agriculture Organization of the United Nations (FAO) (2012) Plant production and protection division: grasslands, rangelands and forage crops. FAO, Rome. http://www.fao.org/agriculture/crops/core-themes/theme/spi/grasslands-rangelands-and-forage-crops/it/. Accessed 20 Oct 2012
3. Van Soest PJ (1994) Nutritional ecology of the ruminant. Comstock Publishing Associates, Ithaca
4. Beever DE (1993) Ruminal animal production from pastures. Current opportunities and future perspective. In: Baker MJ (ed) Forages for our world. Proceedings of the XVII international grasslands congress, SIR Wellington, pp 158–164
5. Food and Agriculture Organization of the United Nations (FAO) (2002) World agriculture towards 2015/2030. FAO, Rome
6. Russell J, Loy D, Anderson J et al (2011) Potential of chemically treated corn stover and modified distillers grains as a partial replacement for corn grain in feedlot diets. A. S. Leaflet R2586, Iowa State University Animal Industry Report 2011
7. Sewell JR, Berger LL, Nash TG et al (2009) Nutrient digestion and performance by lambs and steers fed thermo chemically treated crop residues. J Anim Sci 87:1024–1033
8. Shreck AL, Schneider CJ, Nuttelman BL et al (2013) Varying proportions and amounts of distillers' grains and alkaline treated forage as substitutes for corn grain in finishing cattle diets. Nebraska Beef Cattle Report 56–57
9. Koch RM, Swiger LA, Chambers D et al (1963) Efficiency of feed use in beef cattle. J Anim Sci 22:486–494
10. Archer JAE, Richardson C, Herd RM et al (1999) Potential for selection to improve efficiency of feed use in beef cattle: a review. Aust J Agric Res 50:247–161
11. Nkrumah JD, Okine DK, Mathison GW et al (2006) Relationships of feedlot feed efficiency, performance, and feeding behavior with metabolic rate, methane production and energy partitioning in beef cattle. J Anim Sci 84:145–153
12. Wilkinson JM (2011) Re-defining efficiency of feed use by livestock. Animal 5(7):1014–1022
13. Morell V (2015) Meat-eaters may speed worldwide species extinction, study warns. Science. https://doi.org/10.1126/science.aad1607
14. Machovina B, Feeley KJ, Ripple WJ (2015) Biodiversity conservation: the key is reducing meat consumption. Sci Total Environ 536:419–431
15. Mark W, Ja Z, Haff PK et al (2015) The anthropocene biosphere. Anthropocene Rev 2 (3):196–219
16. Smithers R (2017) Vast animal-feed crops to satisfy our meat needs are destroying planet. The Guardian. Retrieved 3 November 2017
17. Woodyatt A (2020) Human activity threatens billions of years of evolutionary history, researchers warn. CNN. Retrieved 27 May 2020
18. Karim A, Bhat R (2008) Gelatin alternatives for the food industry: recent developments, challenges and prospects. Trends Food Sci Technol 19(12):644–656
19. Easterbrook C, Maddern G (2008) Porcine and bovine surgical products: Jewish, Muslim, and Hindu perspectives. Arch Surg 143:366–370

20. Mariod AA, Fadul H (2013) Gelatin, source, extraction and industrial applications. Acta Sci Pol Technol Aliment 12(2):135–147
21. Kohn R, Dinneen MM, Russek-Cohen E (2005) Using blood urea nitrogen to predict nitrogen excretion and efficiency of nitrogen utilization in cattle, sheep, goats, horses, pigs, and rats
22. FAO (2006) Livestock's long shadow. Food and Agriculture Organisation, Rome
23. International Organization for Standardization (2006) ISO 14040:2006. In: FAO (ed) http://www.iso.org/iso/iso_catalogue/catalogue_tc/catalogue_detail.htm?csnumber=3745. Accessed 2 Apr 2009
24. Jungbluth T, Hartung E, Brose G (2001) Greenhouse gas emissions from animal houses and manure stores. Nutr Cycl Agroecosyst 60:133–145
25. FAO (2010) The global forest resources assessment 2010 – key findings. FAO, Rome
26. IPCC (1997) Intergovernmental panel on climate change/organization for economic cooperation and development: guidelines for national greenhouse gas inventories. Paris. http://www.ipcc-nggip.iges.or.jp/public/mtdocs/pdfles/rockhamp.pdf
27. Mosier A, Kroeze C, Nevison C et al (1998) Closing the global N2Obudget: nitrous oxide emissions through the agricultural nitrogen cycle: OEDC/IPCC/IEA phase II development of IPCC guideline for national greenhouse gas methodology. Nutr Cycl Agroecosyst 52:225–248
28. Steinfeld H, Gerber P, Wassenaar T et al (2006) Livestock's long shadow. Environmental issues and options. Livestock, environment, and development initiative. United Nations Food and Agriculture Organization, Rome
29. Goldewijk K, Beusen A, Janssen P (2010) Long-term dynamic modeling of global population and built-up area in a spatially explicit way: HYDE 3.1. The Holocene 20:1–9
30. Claassen R, Carriazo F, Ueda K (2010) OECD agri-environmental indicators: lessons learned and future directions, U.S. Department of Agriculture–Economic Research Service, Washington, D.C., 23–26 March
31. Devendra C, Thomas D (2002) Crop–animal interactions in mixed farming systems in Asia. Agric Syst 71:27–40
32. Wilson AD, Hodgkinson KC, Noble JC (1988) Vegetation attributes and their application to the management of Australian rangelands. In: Tueller PT (ed) Vegetation science applications for rangeland analysis and management, pp 253–294
33. Conelly WT (1998) Colonial era livestock development policy: introduction of improved dairy cattle in high potential farming areas of Kenya. World Dev 26:1733–1748
34. Nutrient Requirement of Beef Cattle (2000) Seventh revised edition, The National Academies Press (NAP)
35. Stein HH, Berger LL, Drackley JK et al (2008) Nutritional properties and feeding values of soybeans and their co-products. Soybeans chemistry, production, processing, and utilization. AOCS Press, Urbana, pp 613–660
36. Vishnuraj MR, Kandeepan G, Rao KH et al (2016) Occurrence, public health hazards and detection methods of antibiotic residues in foods of animal origin: a comprehensive review. Cogent Food Agric 2:1
37. Marshall BM, Levy SB (2011) Food animals and antimicrobials: impacts on human health. Clin Microbiol Rev 24:718–733
38. Stokestad ELR, Jukes TH (1950) Further observations on the "animal protein factor". Proc Soc Exp Biol Med 73:523–528
39. Gorbach SL (2001) Antimicrobial use in animal feed—time to stop. N Engl J Med 345:1202–1203
40. Howells CH, Joynson DH (1975) Possible role of animal feeding stuffs in spread of antibiotic-resistant intestinal coliforms. Lancet:156–157
41. Levy SB (2002) The antibiotic paradox: how the misuse of antibiotics destroys their curative powers, 2nd edn. Perseus Publishing, Cambridge
42. Smith HW, Crabb WE (1957) The effect of the continuous administration of diets containing low levels of tetracyclines on the incidence of drug-resistant Bacterium coli in the faeces of pigs

and chickens: the sensitivity of the Bact. coli to other chemotherapeutic agents. Vet Rec 69:24–30
43. Feinman SE, Matheson III JC (1978) Draft environmental impact statement subtherapeutic antibacterial agents in animal feeds. Bureau of Veterinary Medicine. Food and Drug Administration, Rockville
44. Halling-Sørensen B, Nielsen SN, Lanzky PF et al (1998) Occurrence, fate and effects of pharmaceutical substances in the environment - a review. Chemosphere 36:357–393
45. Boxall ABA, Fogg LA, Blackwell PA et al (2004) Veterinary medicines in the environment. Rev Environ Contam Toxicol 180:1–91
46. Huang CH, Renew JE, Smeby KL et al (2001) Assessment of potential antibiotic contaminants in water and preliminary occurrence analysis. In: Proceedings of the 2nd international conference on pharmaceuticals and endocrine disrupting chemicals in water, Minneapolis. October 2001. National Groundwater Association
47. Doi AM, Stoskopf MK (2000) The kinetics of oxytetracycline degradation in deionized water under varying temperature, pH, light, substrate, and organic matter. J Aquat Anim Health 12:246–253
48. Halling-Sørensen B (2000) Algal toxicity of antibacterial agent used in intensive farming. Chemosphere 40(7):731–739
49. Tolls J (2001) Sorption of veterinary pharmaceuticals in soils: a review. Environ Sci Technol 35:3397–3406
50. Pinck LA, Holton WF, Allison FE (1961) Antibiotics in soils: 1. Physico-chemical studies of antibiotic-clay complexes. Soil Sci 91:22–28
51. Pinck LA, Holton WF, Allison FE (1961) Antibiotics in soils: II. Extent and mechanism of release. Soil Sci 91:94–99
52. Sithole BB, Guy RD (1987) Models for tetracycline in aquatic environments. I. Interaction with bentonite clay systems. Water Air Soil Pollut 32:303–314
53. Sithole BB, Guy RD (1987) Models for tetracycline in aquatic environments. II. Interaction with humic substances. Water Air Soil Pollut 32:315–321
54. Allaire SE, Del Castillo J, Juneau V (2006) Sorption kinetics of chlortetracycline and tylosin on sandy loam and heavy clay soils. J Environ Qual 35:969–972
55. Krapac IG, Koike S, Meyer MT et al (2004) Long-term monitoring of the occurrence of antibiotic residues and antibiotic resistance genes in groundwater near swine confinement facilities. In: Proceedings of the 4th international conference on pharmaceuticals and endocrine disrupting chemicals in water, Minneapolis. October 2004. National Groundwater Association, pp 158–172
56. Rabolle M, Spliid NH (2000) Sorption and mobility of metronidazole, olaquindox, oxytetracycline, and tylosin in soil. Chemosphere 40:715–722
57. Campagnolo ER, Hohnson KR, Karpati A et al (2002) Antimicrobial residues in animal water and water resources proximal to large-scale swine and poultry feeding operations. Sci Total Environ 299:89–95
58. Kolpin DW, Furlong ET, Meyer MT et al (2002) Pharmaceuticals, hormones, and other organic wastewater contaminants in U.S. Streams, 1999–2000: a national reconnaissance. Environ Sci Technol 36:1202–1211
59. Yang S, Carlson K (2003) Evolution of antibiotic occurrence in a river through pristine, urban, and agricultural landscapes. Water Res 37:4645–4656
60. Chee-Sanford JC, Mackie RI, Koike S et al (2009) Fate and transport of antibiotic residues and antibiotic resistance genes following land application of manure waste. J Environ Qual 38:1086–1108
61. Donohoe AL (1984) Biochemical studies on the fate of monensin in animals and in the 1276 environment. J Anim Sci 58:1528–1539
62. Gavalchin J, Katz SE (1994) The persistence of fecal-borne antibiotics in soil. J AOAC Int 77:481–485

63. Loke ML, Tjørnelund J, Halling-Sørensen B (2002) Determination of the distribution coefficient (logKd) of oxytetracycline, tylosin A, olaquindox and metronidazole in manure. Chemosphere 48:351–361
64. Kolz AC, Ong SK, Moorman TB (2005) Sorption of tylosin onto swine manure. Chemosphere 60:284–289
65. Kolz AC, Moorman TB, Ong SK et al (2005) Degradation and metabolite production of tylosin in anaerobic and aerobic swine-manure lagoons. Water Environ Res 77:49–56
66. O'Neill J (2015) Rapid diagnostics stopping the unnecessary use of antibiotics. http://www.sfam.org.uk/download.cfm?docid=1191C802-F9CA-4406-8DA9ED9E3F7B2A7F.2015.10. Accessed 6 Apr 2018
67. Lim SJ, Seo CK, Kim TH et al (2013) Occurrence and ecological hazard assessment of selected veterinary medicines in livestock wastewater treatment plants. J Environ Sci Health B 48:658–670
68. Watanabe N, Bergamaschi BA, Loftin KA et al (2010) Use and environmental occurrence of antibiotics in free stall dairy farms with manured forage fields. Environ Sci Technol 44:6591–6600
69. Zhu Y, Johnson TA, Su J et al (2013) Diverse and abundant antibiotic resistance genes in Chinese swine farms. Proc Natl Acad Sci USA 110:3435–3440
70. Sarmah AK, Meyer MT, Boxall AB (2006) A global perspective on the use, sales, exposure pathways, occurrence, fate and effects of veterinary antibiotics (vas) in the environment. Chemosphere 65:725–759
71. Underwood JC, Harvey RW, Metge DW et al (2011) Effects of the antimicrobial sulfamethoxazole on groundwater bacterial enrichment. Environ Sci Technol 45:3096–3101
72. Jia S, Zhang XX, Miao Y et al (2017) Fate of antibiotic resistance genes and their associations with bacterial community in livestock breeding wastewater and its receiving river water. Water Res 124:259–268
73. Kennedy D (2013) Time to deal with antibiotics. Science 342:777
74. Levy SB, FitzGerald GB, Macone AB (1976) Spread of antibiotic resistant plasmids from chicken to chicken and from chicken to man. Nature 260:40–42
75. Andremont A (2003) Commensal flora may play key role in spreading antibiotic resistance we need to learn more about commensal flora if we are to better manage this particular window of vulnerability to antibiotic resistance. ASM News 69:601–607
76. Boyazoglu J, Hatziminaoglou I, Morand-Fehr P (2005) The role of the goat in society: past, present and perspectives for the future. Small Rumin Res 60:13–23
77. De Rancourt M, Fois N, Lavín MP, Tchakérian E, Vallerand F (2006) Mediterranean sheep and goats production: an uncertain future. Small Rumin Res 62:167–179
78. Gellrich M, Baur P, Koch B et al (2007) Agricultural land abandonment and natural forest re-growth in the Swiss mountain: a spatially explicit economic analysis. Agric Ecosyst Environ 118:93–108
79. Cocca G, Sturaro E, Gallo L et al (2012) Is the abandonment of traditional livestock farming systems the main driver of mountain landscape change in Alpine areas? Land Use Policy 29:878–886
80. Henkin Z, Ungar ED, Dvash L et al (2011) Effects of cattle grazing on herbage quality in a herbaceous Mediterranean rangeland. Grass Forage Sci 66:516–525
81. FAO (2012) Impact of animal nutrition on animal welfare. Expert Consultation, 26–30 September 2011, FAO Headquarters, Rome. FAO Animal Production and Health Report, No. 1
82. da Silva CS, van den Borne JJGC, Gerrits WJJ et al (2012) Effects of dietary fibers with different physicochemical properties on feeding motivation in adult female pigs. Physiol Behav 107:218–230
83. da Silva CS, Bolhuis JE, Gerrits WJJ et al (2013) Effects of dietary fibers with different fermentation characteristics on feeding motivation in adult female pigs. Physiol Behav 110–111:148–157

84. Bernues A, Ruiz R, Olaizola A et al (2011) Sustainability of pasture based livestock farming systems in the European Mediterranean context: synergies and trade-offs. Livestock Sci 139:44–57
85. Foyer CH, Lam HM, Nguyen HT et al (2016) Neglecting legumes has compromised human health and sustainable food production. Nat Plants 2:16112
86. FAO (2016) Climate change and food security: risks and responses. Rome

7 Conceptual Development of Livestock Supply Chain Management

7.1 Introduction

History of supply chain management is more than a century when it was used to be highly labour intensive, but simple in practice. With the increase in modern transport technology, the concept of supply chain management has been changed with compressive packaging technology, real-time computational network, and cold logistic system with the facility of instant information on goods under transport process. In this connection, both industrial engineering and operational research play significant role. The concept of operational system was initiated during World War II when scientists demonstrated the value of analytics in the military logistics problems. During the 1940s and 1950s, lots of efforts were made to minimize labour intensive practice in material handling and better warehouse system for the safety and security of materials. In the mid-1950s, the concept of transportation management with the development of intermodal containers together with ship, trains, and trucks to handle these containers was initiated for quickening the process of material transport. Soon after a decade, by the 1960s, industrialists started to prefer freight transportation rather than truck and rail. Earlier to the 1960s, all types of documentation were used to be managed manually. The computerization of this data opened new chapter in supply chain management and logistics planning, from randomized storage in warehouses to optimization of inventory and truck routing. In the late 1970s and early 1980s, many premier institutes like the Georgia Tech of the Production and Distribution Research Centre, the Material Handling Research Centre, and the Computational Optimization Centre were involved in to find out the applied aspects of theoretical model on microchips based supply chain management process.

During the 1980s, logistics application in supply chain management was widely accepted by industries. The emergence of personal computer brought tremendous boost in planning strategies for developing flexible spreadsheets and map based interfaces which enabled remarkable developments in logistics planning and execution technology. This period has witnessed the complete systematic process of

P. K. Rout, B. K. Behera, *Sustainability in Ruminant Livestock*,
https://doi.org/10.1007/978-981-33-4343-6_7

feedstock procuring methods, manufacturing process management, and logistic aspects of products transfer to end users safely and timely. The manufacturers started thinking first on product quality, quality assurance and quality management, efficient logistic process for transferring products from site of manufacturing to end users before planning the development of an industry.

The industries started realizing the importance of logistics in developing the efficacy of material transition. The 1990s faced the technology revolution in the area of supply chain management and logistics network development mainly due to material requirement planning system. In the beginning of the 1970s, the promoters started to minimize inventories quality in order to reduce the cost. In this connection, the manufacturer started introducing material requirement planning (MRP) and manufacturing resource planning (MRPII) systems to minimize inventory holding. In due course of time, the effective materials management gained appreciation. This could only be possible by the application of highly sophisticated software packages (LAN and WAN) related to information technology. The manufacturers started tracking the goods in logistic process to understand inventories condition. Digital process like JIT and TQM became an integral and unavoidable method for getting update information on goods transition, reaming at manufacturer administrative desk. Due to continuous increase in competitive market, tremendous pressure on world economy was noticed. The complete systematic process of feedstock procuring methods, manufacturing process management, and logistic aspects of products transfer to end users should be safe and timely. The manufacturers started thinking first on product quality, quality assurance and quality management, efficient logistic process for transferring products from site of manufacturing to end users before planning the development of an industry. Besides, traditional trade flow and the globalization of production became a catalyst in developing complex market scenario and issue on labour intensification. The industrialized developing countries started facing challenge for globalization of products due to heavy competition with industrial sector of developed nations.

7.2 Livestock Products Supply Chain Management

Livestock (sheep, goat, pork, cattle, and beef) marketing and supply chain used to be restricted in a specific locality (Fig. 7.1). Due to perishable nature of meat, the small-scale livestock producers used to face a number of challenges like restricted access to markets and related services, environmental constraints, and limited capacity. Besides, even, selling the live animal was used to be tough due to lack of proper transport system. Modernization of supply chain management has entirely changed the scenario of global livestock milk (Fig. 7.2) and meat (Fig. 7.3) transport and marketing process.

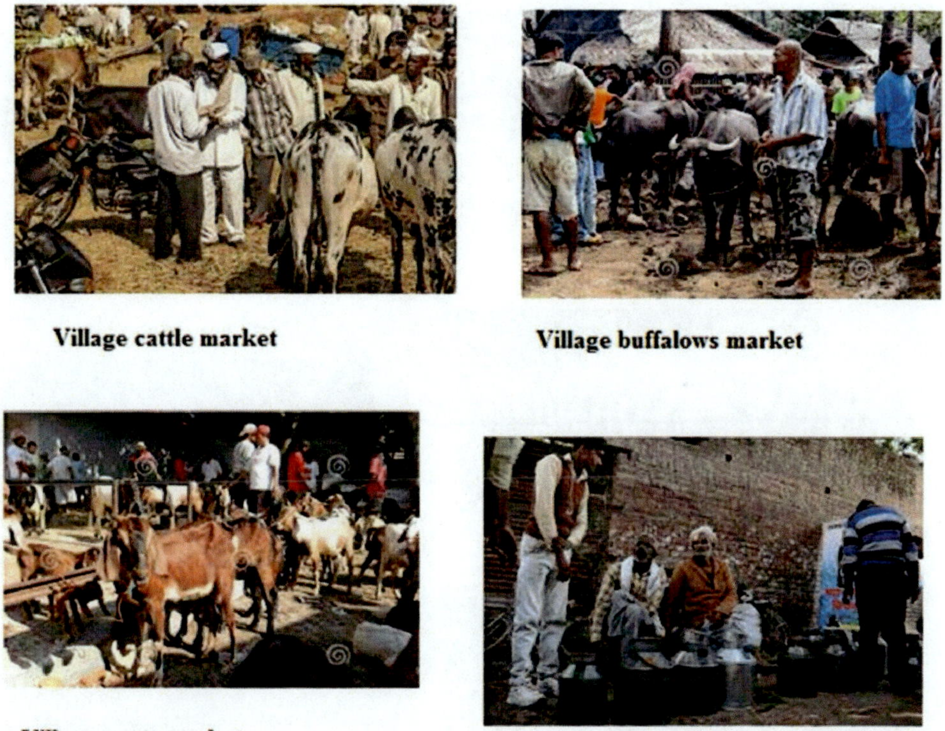

Village cattle market **Village buffalows market**

Village goats market **Village milk selling**

Fig. 7.1 Selling of livestock and products (milk and meat) in and around the village

7.2.1 Milk Supply Chain Management

Cattle, buffalo, and goats are grown under agricultural practice. Production of food energy and protein from dairy cattle systems is more efficient than from beef producing systems [1–3]. Cattle and buffalo are major contributors of milk in the world. Non-bovine milk such as goat, sheep, and camel is also popular. Goat and cattle are having high demand for milk in Africa and parts of Southeast Asia, whereas in India and Pakistan buffalo milk is also popular. The average volume of cattle milk produced worldwide has been rising steadily. In 2015, the cattle milk production was about 497 million metric tons, and it was raised to 522 million metric tons, by the end of 2019. India is the largest producer of milk followed by United States of America (USA), China, Pakistan, and Brazil. The buffalo milk is produced in the Southeast Asian region, India, and Pakistan.

In order to promote social inclusion and women empowerment, for the first time in the history of world, in 1948, Gujarat Co-operative Milk Marketing Federation Ltd. (GCMMF) was formed, which today is jointly owned by 3.6 million milk producers in Gujarat. Anand Milk Union Limited or Amul is an Indian dairy cooperative based at Anand in the State Gujarat for collecting, manufacturing, and

Fig. 7.2 Milk supply chain management

selling milk both at retail dealers and wholesale dealers all over India, and gives an excellent concept on milk supply chain management system [1]. It is a branch of GCMMF.

Cow, buffalo, and goat milk are the major milk. In less-developing countries goat milk demand comes under third option, after cow and buffalo. After cattle and

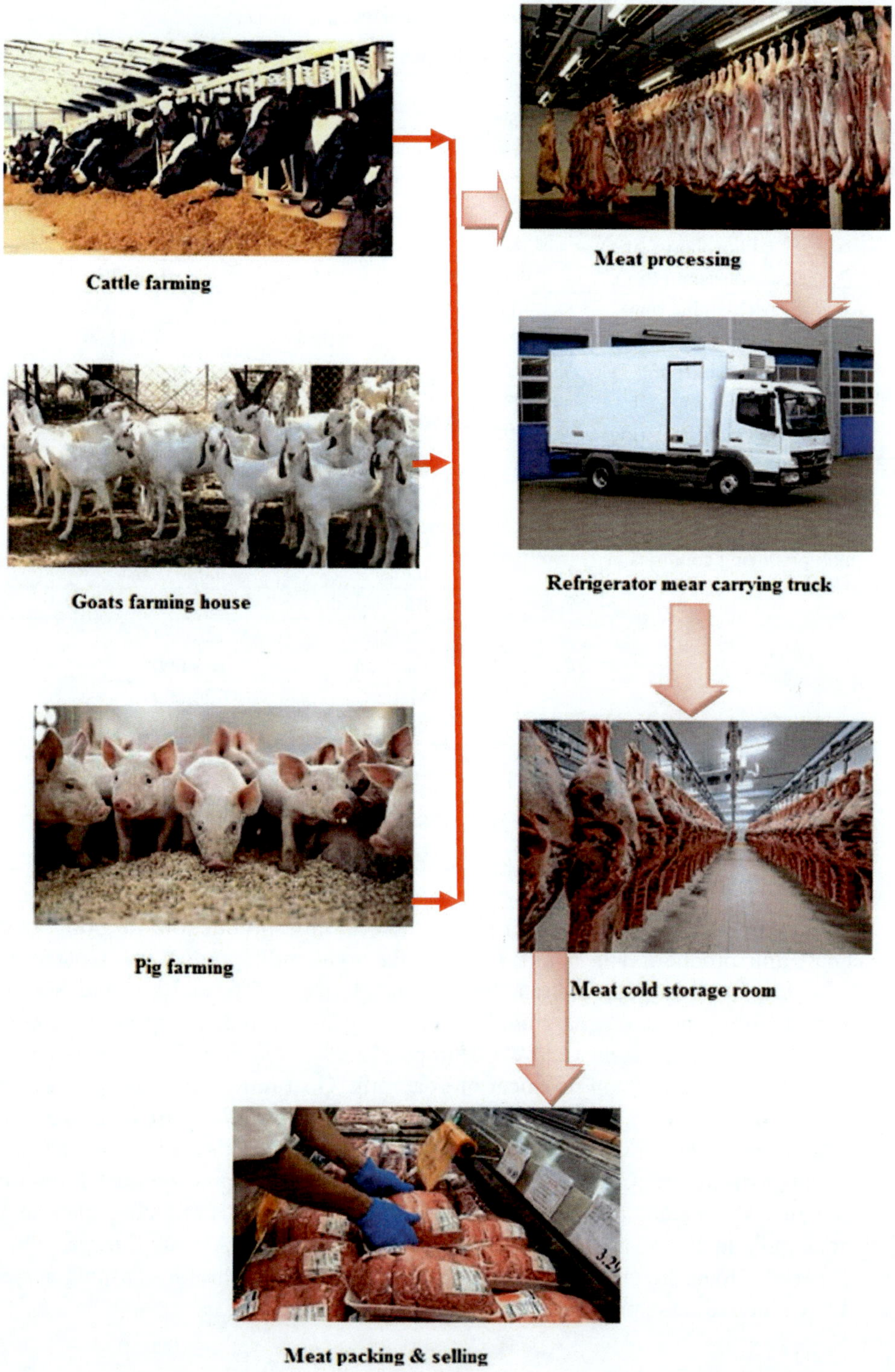

Fig. 7.3 Livestock meat supply chain management

Table 7.1 Comparison of milk properties of goat, cow, and human

No	Compound	Goat (250 mL)	Cow (250 mL)	Human (250 mL)
1	Energy (Kcal)	168	146	172
2	Protein (g)	8.69	7.86	2.53
3	Fat (g)	10.10	7.95	10.77
4	Cholesterol (mg)	27	24	34
5	Calcium (mg)	327	222	79
6	Phosphorus (mg)	271	222	34
7	Potassium (mg)	498	349	125
8	Thiamine (mg)	0.117	0.107	0.034
9	Riboflavin (mg)	0.337	0.447	0.089
10	Niacin (mg)	0.676	0.261	0.423
11	Vitamin A (IU)	483	249	522
12	Tryptophan (mg)	0.107	0.183	0.042

Source USDA National Nutrient database for standard reference

Table 7.2 The top goat milk producing countries in the world

Rank	Country	Production (Metric Tonnes)
1	India	5,000,000
2	Bangladesh	2,616,000
3	Sudan	1,532,000
4	Pakistan	801,000
5	Mali	720,000
6	France	580,694
7	Spain	471,999
8	Turkey	415,743
9	Somalia	400,000
10	Greece	340,000

buffalo, goat contributes a major share to world milk production. In global market goat milk production is about 1.5% of the total milk production. Goat milk is produced in all regions of lran, Iraq, Syria, Africa, the Near East, and Southeast Asia. In Southeast Asia goat milk is produced mainly in India and Bangladesh. Four continents, namely Asia (57.8%), Europe (13.7%), Africa (25.2%), and America (3.2%) are globally main producer of goat milk. Goat milk is also as good as cow's milk, and it can be well supplemented in place of cow's milk where its availability is in scarcity (Table 7.1). India, China, Bangladesh, and Pakistan keep 46.7% of the world goat livestock, and 55.60% of the world goat milk produced by India, Bangladesh, Sudan, and Pakistan (Table 7.2). The most producing continent of goat milk in the world is Asia, which is followed by Africa, Europe, and the America. Goats are small ruminants that fit well into smallholder farming systems. They required less space than larger ruminants.

7.2.2 Meat Supply Chain Management

The global beef market size is expected to capture 383.5 billion, by the end of 2025. Beef and veal contain high protein as compared to chicken, pork, turkey, and lamb. Surging consciousness about high nutritional value of animal protein has been gaining high market demand. In the USA, people prefer to eat beef over other meat. Chinese people also prefer to take beef rather than pork or chicken. Due to increase in beef demand, the government of China lifted the ban on Australian chilled frozen food.

USA is one of the largest producers of beef in the global market. Demand from restaurant chains such as Burger King and McDonald's is also expected to contribute to the growth of the US market. In addition rise in halal beef demand by Islamic population is also responsible for popular beef market. Modern technology related to slaughtering techniques and improvements in animal feed technology have increased the confidence level of animal protein lovers and demand for animal meat.

The modern meat supply chain management process provides consumers the best quality of meat which is also one of the major factors for increase in demand of animal proteins (Fig. 7.3). The main target of supply chain management is to develop confidence in consumers on safety and security of quality meat, timely.

Goats and sheep are small ruminants, and can be grown under available natural conditions with minimum management and less expenses. The livestock owner sells goats to butchers for meat processing and marketing on the basis of customers demand from time to time. The wholesale dealer sells the carcasses or meat cuts to retail dealers with proper hygienically packages to consumers on a demand basis. Goat meat is known as *Chevon* (*Fr.*) and *capretto* (*Sp.*), or *kid* when is from young animals. The word mutton is used for sheep meat in some part of Asia (India, Pakistan, Nepal, Sri Lanka, and Bangladesh). Goat meat is consumed in major parts of the world. The developed countries enjoy goat meat as exotic livestock, whereas in developing nations like Southeast Asia and Africa goat is the major source of meat production [4].

The small ruminant marketing is complex and it is having inefficient marketing chain due to participation of various players in developing countries. Primarily, the difficulty is to access marketing chain, and this is mainly due to inadequate access to market outlets, poor marketing management concept, and lack of awareness of small farmers. Thus, goat production and marketing are messy without quality control and consistency in animal breed, body weights, and health condition at slaughter house. Additionally, goat supply is not properly managed on the basis of consumer demand across time. This is due to wide fluctuations in prices paid to producers and paid by consumers which ultimately lead to disheartenment of the producer to improve in quality supply of goats in the market. The geographical location variability in goat production and availability of suitable facilities for goat slaughtering and processing become the primary cause of marketing cost.

The demand of goat meat at global market can be improved by giving value addition and quality to goat meat. This can be well managed by improving the breeding quality of goat and environmental factors, particularly better nutrition and health care. By this practice, the consistency in goat body weight over time can be well managed to attract the attention of buyers. The goat meat sales value can be increased by selling goat carcasses into cuts on demand basis. Deboning goat carcasses is labour intensive and expensive, but it improves the quality of meat for making muscles into sausages, bolognas, stix, jerky, etc.

7.2.3 Therapeutic Proteins from Livestock

The dawn of therapeutic protein production for human health care was initiated with microbes like *Escherichia coli* and *Saccharomyces cerevisiae* as host cells for the production of simple polypeptides such as insulin and human growth hormone. But the microbes based bioreactor has certain limitation as microbial model systems are not suitable for complex posttranslational modifications or intricate folding requirements, such as the coagulation factors or monoclonal antibodies. Industrialization of recombinant production of complex molecules, such as antithrombin and 1-antitrypsin, has not yet been reported by using microbial cells or mammalian cells as host systems, in a bioreactor. Currently, human plasma is only the alternative source for antithrombin and 1-antitrypsin, both quality and quantity wise. During the late 1980s, reports on successful expression of therapeutic proteins in milk of genetically engineered animals like sheep, goat, pig, and cow have created new history in biopharmaceutical sciences [5–7].

While moving from microbes based bioreactor to variety of host cells (animal tissues, plant tissues, insects, and CHO) based bioreactor, it has been observed that transgenic livestock animals like sheep, rabbits, goat, and pig are the best sources for producing recombinant therapeutic proteins [8] because the mammary gland is the best choice of expression of recombinant therapeutic proteins. Interestingly, the past literature showed that 100,000 kg of human serum albumin can be produced from 5400 cows, which can meet the demand of 100,000 kg of human serum albumin per year, globally. Five thousand kg of antitrypsin (α-AT) production can be achieved by 4500 sheep, and 100 kg of monoclonal antibodies (mAbs) will be produced from 100 goats, and 75 kg of antithrombin III can be produced from 75 goats and only two pigs can be a good source of 2 kg of human clotting factors IX [9].

Thus, the development of therapeutic proteins technology from microbes to mammary gland bioreactor has proven to be a better option for human protein expression (Fig. 7.4). Interestingly, it has been observed that the goat milk contains several types of human proteins (Table 7.3).

Thus, especially genetically modified livestock animal (cows and goats) can be used for the production of therapeutic recombinant proteins due to production of adequate milk with minimum maintenance expenditure (Table 7.4). The state-of-the-art of production of transgenic goat has diverted attention of pharmaceutical industrialists for therapeutic proteins production. The genetically engineered goat is supposed to be a better alternative to small ruminant for the use as bioreactor for the production of recombinant therapeutic proteins in their mammary gland (Fig. 7.5).

Fig. 7.4 GEA Process Engineering in New Zealand has built a complete process plant for infant formula and goat milk powder for its long-term customer Dairy Goat Co-operative (DGC) in Hamilton. Working closely with DGC, GEA has used its experience in plant design to develop a highly innovative plant that provides multiple benefits including: improved product quality, reduced energy costs, and reduced water consumption

Table 7.3 Composition of different types of therapeutic human proteins in goat milk

Gene (Source)	Vector	Promoter	Expression (mg/mL)	References
Antithrombin III (human)	Adenovirus	CMV	2.8	Yang et al. [10]
Growth hormone (human)	Adenovirus	CMV	2.4	Han et al. [11]
Erythropoietin (human)	Adenovirus	CMV	2.0	Toledo et al. [12]
Lactoferrin (human)	Adenovirus	CMV	2.0	Han et al. [13]
E2-CSFV (swine)	Adenovirus	CMV	1.2	Toledo et al. [14]
Growth hormone (human)	Adenovirus	CMV	0.3	Sánchez et al. [15]
NGF- β (human)	Adenovirus	CMV	0.2	Xiao et al. [16]
Lysostaphin (bacterian)	Adenovirus	RSV	0.02	Fan et al. [17]
Lysostaphin (bacterian)	Adenovirus	RSV	0.001	Fan et al. [18]
Growth hormone (human)	Retrovirus	MoMLV	6.0×10^{-5}	Archer et al. [19]
Factor IX (human)	Retrovirus	β-actina	4.0×10^{-5}	Zhang et al. [20]

The first human biological drug (antithrombin-AT) was approved by the European Agency of the Evaluation of Medical Products and, after in the USA, by Food and Drug Administration [36]. The goat produces about sevenfold to eightfold increase in the amount of serum from rabbit. Besides this, the goat serum contains

Table 7.4 List of therapeutic protein produced from transgenic goat milk

Type of therapeutic proteins	Purpose	References
α-fetoprotein (hαFP) (serum glycoprotein)	Autoimmune diseases such as rheumatoid arthritis, multiple sclerosis, myasthenia gravis, and psoriasis	Parker et al. [21]
Malaria vaccine antigen (expression vector for the antigen $MSP1_{42}$ of malaria) non-glycosylated form produced milk over 100 days capable of generating the equivalent of five million doses of vaccine against malaria.	For malaria prevention	Behboodi et al. [22]
Antithrombin III (hAT-III) (anti-inflammatory and anticoagulant properties) approved for clinical use in the European market (Schmidt, 2006), and in 2009 in the USA (Kling, 2009)	Anti-inflammatory and anticoagulant properties (hAT-III inhibits the majority of the proteinases in blood coagulation, such as thrombin, activated factor X, and, to a lesser degree, the activated coagulation factors, including IXa, XIa, XIIa, urokinase, trypsin, plasmin and kallikrein)	Schmidt [23], Kling [24], Edmunds et al. [25]
Human tissue plasminogen activator (htPA)	Capable of initiating the degradation of proteins of the extracellular matrix	Ebert et al. [26], Baldassarre et al. [27]
Butyrylcholinesterase (hBChE) (Human plasma)	Cholinergic neurotransmission and may be involved in other functions of the nervous system and neurodegenerative diseases treatment of humans exposed to organophosphorus or chemical warfare agents (Lenz et al., 2010)	Huang et al. [28]
Lactoferrin (LTF)	Antiviral, antitumor, antibacterial, antifungal, anti-inflammatory, and immunoregulatory properties (Varadhacharyet al., 2004)	Zhang et al. [29]
factor IX (hF IX)	Haemorrhagic disorders of Haemophilia B	Huang et al. [30]
Lysozyme	Antimicrobial properties and catalyses the cleavage of glycosidic bonds between the C-1 of N-acetylmuramic acid (Mur2Ac) and the C-4 of N-acetyl-D-glucosamine (GlcNAc) in peptidoglycans of bacterial cell walls (Maga *et al.*, 2006b)	Maga et al. [31]
Human Granulocyte Colony-Stimulating Factor (hG-CSF)	Widely utilized in different forms of neutropenia, chemotherapeutically induced leukopenia and mobilization of progenitor cells for autologous or allogenic transplants	Ko et al. [32], Freitas et al. [33, 34]

(continued)

Table 7.4 (continued)

Type of therapeutic proteins	Purpose	References
Human growth hormone (hGH) or somatotropin	One of the principal hormones required for post-natal growth and is absolutely essential for normal body development	Lee et al. [35]

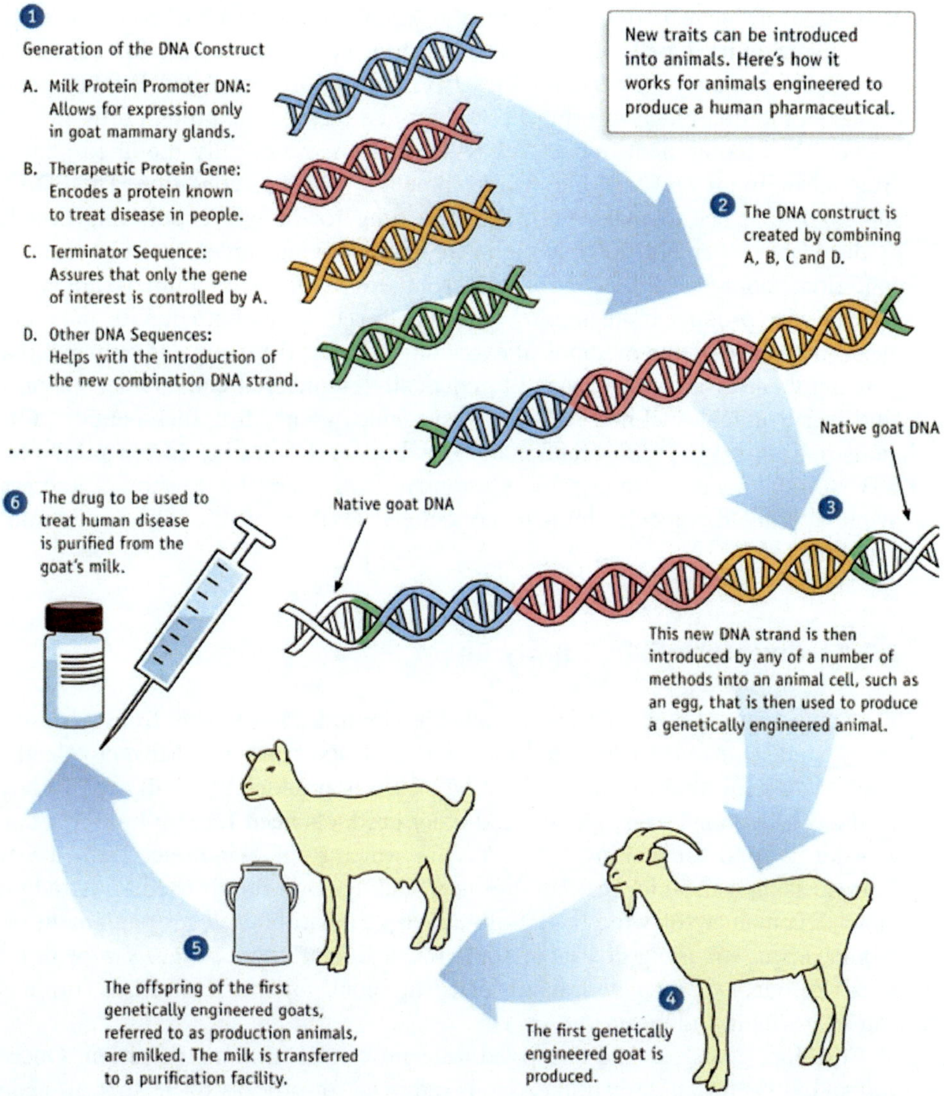

Fig. 7.5 Genetically engineered goat

about 20 mg/ml of total IgG, which is about 2–3 times more than rabbit serum. Thus, goat model bioreactor is a more cost-effective alternative for producing large amount of antibody in a highly commercialized competitive market. The most important fact is getting antibody from single animal, rather than several different individuals.

The global therapeutic proteins market was about $140,109 million, and it is predicted that at the end of 2023 the market size will be $ 217.23, registering a CAGR of 6.5% from 2017 to 2023. This is highly encouraging to pursue commercialization of "goat model" as bio-reactor to compete with other competitors presently using animal cells, plant cells, bacteria, etc. as host cell for recombinant therapeutic proteins. rEVO Biologics (formerly known as GTC Biotherapeutics) has taken the first credit to bring FDA approved therapeutic protein drug in the US market, the recombinant human AT (ATryn) from genetically modified goat. The drug, meant to prevent fatal blood clots in people with rare condition. The FDA also approved goat use to make the drug. The drug for human anticlotting protein is produced by a herd of 200 bioengineered goats living under carefully controlled conditions on a farm in central Massachusetts. The farm animals are a better alternative to produce biotechnology drugs at lower cost by both quality and quantity means than the existing methods of extracting proteins from donated human blood or growing them in large steel vats of genetically engineered cells. Other nations are also very much inclined to use transgenic goats for therapeutic proteins industrialization. ATryn (brand name), a human IX (hFix) factor produces by CTC Biotherapeutic, Inc. [37]. Currently, promoters have shown interest in manufacturing therapeutic human monoclonal antibodies through goat mammary gland [38].

7.2.4 Livestock Skin Supply Chain

The natural skin of cow with hair is called as cowhide. The leather from it is used for shoes, wallets, leather jacket, furniture, and car leather and belt. Most cow leather is available without hair. Like cowhide, goatskin is produced globally for variety of products like shoes, bags, gloves, and other products need for soft hide. Un-tanned goatskin is used for making containers for water, kefir, wine, etc. Tanned leather from goatskin is used for making rugs and binding. Goatskin is used for a traditional Spanish container for wine bag. Non-tanned goatskin is used for parchment or for drumheads or sounding boards of some musical instruments, e.g. *misnice* in mediaeval Europe, *bodhran* in Ireland, *esraj* in India, and for instrument drum skin named *bedug* in Indonesia.

Cowhide, goatskin, and sheep skin are processed in similar fashion. Once the animal has been slaughtered, the skin is removed for process for leather. In general, the leather manufacturing process is divided into three steps: preparatory stages, tanning, and crusting. For fine finishing, the leather is subjected to sub-processing, which may not be required for all type of leather. In preparatory stages, the raw hide is subjected to salt treatment for removal of hair and unwanted flesh. This process includes soaking, hair removal, liming, deliming, bating, bleaching, and pickling.

Tanning process is mainly carried out for the stabilization of collagen of the raw hide to enhance thermal, chemical, and microbiological stability of the hides and skin, making it suitable for a wide variety of end applications. The raw hides, in dry form, becomes hard and inflexible. But the tanned hide dries to a flexible form that does not become putrid when rewetted. In tanning process the hides obtained from preparatory stage are immersed into a drum containing the tanning "liquor". The hides soak while the drum slowly rotates about its axis, and allow the tanning liquor to penetrate into the skin, slowly. Once, the hide becomes saturated with tanning liquor, the pH is raised slowly to bring it to basification for fixing the tanning materials to the leather. The hydrothermal property and shrinkage nature are mainly depending on the basification of the hide. Crusting is the final process to give finishing touch to leather like thinning and developing lubricate property of leather. On the basis of market demand, the leather is coloured on the basis of customer's requirement, crusting culminates with a drying and softening operation, and may include splitting, shaving, dyeing, whitening or other method. Besides, the finishing also includes oiling, brushing, buffing, coating, polishing, embossing, glazing, and tumbling.

7.3 Livestock and Cold Chain Logistics

At present, refrigerated and frozen logistics cost is about $13.4 billion/year with the annually growing rate of 5–6%. It has also been assumed that this value may change to 8–9% over a period of subsequent years, and may last for a longer time. Some of the livestock (goat and cattle) products need proper cooling. Milk is stored over 4 °C (40 °F) or cooled too slowly. Proper cooling is critical for the maintenance of quality and prevention of any microbial infection. As compared to bovine milk, goat milk is more sensitive to temperature and pH [39–42]. The heat sensitivity of goat milk depends on environmental factors and the practice of farming management. In goat milk, transforming process from ultra high temperature treatment (UHT, 136 °C/6 s) to low temperature needs special care.

Mostly, local cow milk products marketing, with special reference to rural areas, were lack of cold chain logistic process. It is necessary to develop well-trained stakeholders to manage milk transport system with quality security and safety. India is the world's largest milk producer. The Indian dairy sector is different from the other dairy sector of the world. In India, the emphasis has been given on both cattle and buffalo milk.

In spite of the world's highest milk producing country, the farmers face problems in keeping cattle due to urbanization of grazing land and late maturity in most of the cattle breeds. In India, the excessive numbers of unproductive animals create utilization of available feeds and fodder. The ever increasing gap between demand and supply of feeds and fodder to the total requirement has been creating hurdles in milk production. As a result, dairy farmers are not getting remunerative price for milk supply. Due to practice of extensive cross-breeding programme, the fat content of cross-breed cow's milk is on the declining condition and low price is offered as the milk price is estimated on the basis of fat content.

Presently, the problem related to goat milk marketing is also due to inadequate sequential of commercial process technologies involved in logistic process. Generally, the raw milk is thoroughly inspected before its transfer to logistic systems for distribution. Goat milk is an alternative feed source for the dairy goat producer involving in buying and marketing lambs, veal calves, or piglets. So, it is necessary to have additional marketing skill. While marketing fluid goat milk the dairy owner should have proper strategy for getting additional income from the kid goat crop. Generally, kid goats are raised to different market weights, and for different market seasons. At present powdered goat milk is gaining commercial importance due to its more shelf-life without any substantial changes in quality [43, 44]. In addition, powdered goat milk is easy to handle by logistic system for distribution at the cost of low expenditure [45]. The dehydration technology is helpful for the consumers with year-round marketable product.

Spray drying is a common technology for making powered milk of any milk. Powdered caprine milk is used for variety of products like ice cream, chocolate, yogurt, infant formula, and cheese [46]. The global goat milk market is expected to post a CAGR of over 6% during the period 2018–2022.

The main reason for increasing demand of goat milk is their use for production of low lactose or lactose free products. This is required for the people having lactose intolerance, and the number of lactose intolerant people is in increasing order. Goat milk infant formula is also known as breast milk formula. It is a formulated product with the combination of goat milk as raw material. This is marketed to fulfil the requirement of bottle-feeding babies allergic to cow's milk. Dairy Goat Co-operative (DGC) ranks the first in terms of revenue share in global market of goat milk infant formula having 9.99% of market share in 2016, followed by Ausnutria Dairy with a market share of 6.75%. Some of the major key players functioning in the Goat Milk Infant Formula Market Report include: *DGC, Ausnutria Dairy, Baiyue Youlishi, YaTai-Precious, Red Star, Guan Shan, Milk Goat, Herds, Fine boon, Jinniu, Shengfei, ShengTang, Holle, FIT, Vitagermine, Xi Yang, Baiyue Dairy Group.*

7.3.1 Facility Growth

cGMP based highly sophisticated processing unit with integrated operation management, temperature controlled warehouse facilities, and global linked supply chain management are integral part of third-party logistics service. The most important salient feature of third-party logistics service is based on customer's need and market condition, such as demands and delivery process requirement on the basis of product and materials. On the basis of special requirement, the third-party logistics go behind their activities and provide value-added service related to the production and procurement of goods, i.e., services that integrate parts of the supply chain, and when this integration occurs, the provider is then called a third-party supply chain management provider (3PSCM) or supply chain management service provider

(3 PSCM). The 3 PL targets particular fractions within supply management, such as warehousing transportation or raw material provision (www.hexaresearch.com 2018).

It has been observed that global third-party logistics market supposed to gain 5% CAGR during the period 2014 to 2016 with the coverage of shipping industry to supervise logistic undertaking (forecasting, warehousing, and conveyance management software). This market will gain the size of about 1054 billion by the end of 2024 (www.hexaresearch.com 2018). Presently, the cool chain supply management system has become sophisticated with temperature controlling system on manufacturers need base. In addition the provision of real-time tracking provision gives guarantee to the life of product as prescribed by the manufacturer. As a result strong competition is going on between bulk air-cargo container suppliers, such as Envirotainer and CSafe, and suppliers of single-use pallet shippers such as Sonoco ThermoSafe and Cold Chain Technologies.

On the basis of latest technology, some of the multinational groups like Dairy Goat Co-operative *(DGC), Ausnutria Dairy, Baiyue Youlishi, YaTai-Precious, Red Star, Guan Shan, MilkGoat, Herds, Fineboon, Jinniu, Shengfei, ShengTang, Holle, FIT, Vitagermine, Xi Yang Yang, Baiyue Dairy Group have involved in goat milk processing with highly integrated technology with minimum expenditure of water and energy. The DGC plants are having world class facility for manufacturing milk formula, and also monitor the entire process from milking to distribution. With the strong European market DGC grew its revenue to* $ *193 million last season (2016). DGC is specialized in developing, manufacturing, and marketing its own brand of goat milk nutritional power for infants and children. The products are exported to more than 27 countries.*

Generally, essential fatty acids, folic acid, vitamin B12, C, D, E, iron are added to modify goat milk on a requirement basis. When making formula from cow milk, manufacturers add extra processing steps to remove and replace the milk fat, extract or add whey proteins plus ingredients like nucleotides or oligosaccharides. Addition of whey proteins to cow milk is favoured by many formula manufacturers to improve the supply of essential and semi-essential amino acids [47, 48]. However, DGC's research conforms; whey is not necessary for formula made from goat milk. Goat milk only requires the addition of very small amounts of two amino acids to supply all the essential and semi-essential amino acids needed for infants. DGC process for goat milk involves in retaining milk fats before adding lactose, essential fatty acids, vitamins, and minerals. In addition little quantity of vegetable oil is added to top up the essential unsaturated fatty acids in goat milk without removing milk fat [49], because goat milk fat is having the property to retain the medium chain saturated fatty acids which can be easily absorbed by infants than long chain fatty acids [50]. Generally, goat milk is rich in nucleotides compared to cow milk, thus there is no need to add nucleotides while making formula from whole goat milk. DGC has special technique to avoid smell without damaging milk fat [48].

7.3.2 Clinical Logistics and Patent Rights

Due to complex and sequential regulatory steps, it has been an extremely time-consuming process to have FDA approval of biologics derived from different host cells based bioreactor or through animal mammary gland bioreactor. In addition, the logistics systems management involve in carrying to multiple locations is a tough challenge due to the high temperature sensitive biomolecules or animal tissues or plant tissues. Presently, the clinical logistics bear the responsibility of providing laboratory facility services and ancillary supplies like providing cGMP laboratory testing and diagnostic equipment, information bulletins, and other associated facilities. Laboratory services include distribution of patient forms and kits, laboratory sample handling with temperature regulated equipment, computerization and compiling of testing results, and keeping integration of overall services.

The main logic behind the highly effective clinical logistics is to monitor global marketing on the basis of feedback from the end users. This needs the efficacy of timely technical services, innovative temperature controlled warehouse facilities, and safe delivery of testing sample to different geographical locations.

In 2011, GTP Biotherapeutics Inc. (an LFB group company) got patent granted from the United States Patent and Trademark Office (USPTO) on DNA constructs for the production of any therapeutic protein in the milk of any transgenic animal. This "Broad DNA" patent extends to 2027. In addition, GTP received the US patent for ATryn (recombinant human antithrombin). ATryn was the first and still only, trans-genetically produced therapeutic protein approved by both the FDA and EMA authorities. The main claim of this patent is related to a method of treating subjects having an antithrombin III deficiency or inflammation with the administration of recombinant antithrombin III produced in the mammary gland of a non-human mammal.

Activities in goat therapeutic proteins are in progress with other countries like Iran, where human factor IX was already produced, but is still under clinical trial. Although active R&D research is under progress on expression of human monoclonal antibodies in milk of transgenic goats, there are no data on transgenic goat producing monoclonal antibodies except from annual report of GTC-Biotherapeutics, 2010. Expression of human monoclonal antibodies in goats against CD20 receptor, CD 137 receptor, tumour necrosis factors (TNF), and epidermal growth factor receptors (EGFR) is being documented long back [38]. Past reports also support the possibility of production of recombinant hormones, growth factors, and cytokinase, G-CSF [51, 52], EPO lactoferrin [53–55] by using transgenic goats.

7.3.3 Logistics Trends

Like other biologics the cattle and goat milk therapeutic proteins, and clinical testing kits, are highly temperature susceptible, and need intensive care in handling during logistic operation. The protocols followed for biologics are also applicable for handling goat therapeutic proteins. Mostly, the products are shipped at room temperature, after proper packaging, and thereafter pass on to logistic operation systems under temperature controlled condition with the provision of digital tracking devices.

Since more than two decades the bio-pharmaceutical supply chain management under temperature controlled condition has been in progressive form. The introduction of digital IT based operation system and tracking of products during logistics operation condition have not only developed the practice of timely delivery but also taken care of product quality management till it reaches the end users. The major top transgenic goat milk derived therapeutic protein manufacturers like *DGC, Ausnutria Dairy, Baiyue Youlishi, YaTai-Precious, Red Star, Guan Shan, MilkGoat, Herds, Fineboon, Jinniu, Shengfei, ShengTang, Holle, FIT, Vitagermine, Xi Yang Yang, Baiyue Dairy Group have their own logistics systems for transfer of goats milk derived products to different location on demand basis.* In addition, the manufacturers are having well-organized groups for packing, freight forwarding and transportation, and electronic monitoring of both a packing's internal conditions and process followed by carriers, to meet the requirement of industries for sustainable growth and development.

7.3.4 Capacity Addition

Due to continuous increase in patent expiration (including blockbuster drug), the challenged market dynamic and development of awareness on the benefit of biologics, the bio-pharmaceutical industries are in extreme dilemma whether to continue further with the stainless bioreactor based host cells culture and production of therapeutic proteins or go for the transgenic animal bioreactor for the expression of human proteins and other related biomolecules. From the market prospective it is highly necessary to reduce the accelerating price of therapeutic proteins and compete with global competitors. So, the promoters are interested in adapting animal bioreactor for cost-oriented therapeutic protein production. For this additional facility development, a company needs to either develop super specialized R&D or opt for a group having well experience in handling animal bioreactor for therapeutic protein production. The best example for this sort of collaboration programme for the production of quality products on the basis of market demand was initiated by rEVO Biologics originally known as GTC Biotherapeutics Inc. GTC Biotherapeutics Inc. Framingham, earlier known as Genzyme Transgenic Corporation has entered into an agreement of mutual understanding collaboration work with French firm LFB Biotechnologies to design and develop recombinant plasma proteins and monoclonal antibodies by handling knowhow technology of GTC's transgenic animal development facilities for the production of targeted products, and retain exclusive commercial rights to the products in North America. The responsibility of clinical development will be taken by LFB Biotechnologies and will have exclusive commercial rights in Europe. GTC and LFB Biotechnologies will hold co-exclusive right in the rest of the global market. In 2013, GTC Biotherapeutics changed its name to rEVO Biologics for the treatment of rare diseases. rEVO Biologics corporate offices are in Framingham, MA, with protein production facilities in Charlton, MA.

To shift from primary care towards speciality drugs and biologics, the bio-pharmaceutical manufacturers need additional space for sophisticated equipment, space for continuous processing, working place for manpower, and facilities for inventories storing and third-party logistics operation. From the logistics perspective, the value of biological origin products is comparatively insignificant with reference to minerals, automotive parts, agriculture feedstock, other high-volume materials, but from storing and transport points of view special care is needed for sophisticated and expensive machines and cryo-preservative equipment with the provision of dedicated spaces. It has been noticed that FedEx Supply Chain added 1.1 million square-feet multitenant distribution warehouses on East Holmes Road.

UPS Healthcare Logistics, which operates more than 60 facilities for healthcare products worldwide, built a second facility near Bogotá, Colombia in the past year, adding 76,000 sq. ft. there, and bringing its global footprint to more than seven million sq. ft. The company also expanded its Express Critical delivery service, which has been in place in the USA, to Europe. Arch-rival DHL Global Forwarding opened a new Life Sciences Centre of Excellence in Ireland, the 43rd country with such a facility (some countries have multiple centres. The 40,000-sq. ft. facility is close to Dublin Airport, and features room-temperature, refrigerated, and frozen dedicated capacity, along with repackaging services. It has been observed that with the increase in clinical trial of biopharmaceuticals the logistic expenses are in increasing order at global level.

7.3.5 Cold Chain Operation

The act of cold chain management of livestock derived products is similar to other biologics logistics operation process. Due to high susceptibility to temperature, some of the high value added products produced from livestock are delicate to handle in supply chain management (SCM) process. Thus cattle, goat, and sheep milk products and therapeutic proteins SCM is executed with utmost temperature controlling system and well-managed tracking provision. Reducing or eliminating temperature excursions during shipping is a highly responsible task. The small containers holding clinical trial materials are equipped with temperature recording instruments to understand and assure the quality safety of cattle and milk biologics. In addition, it is mandatory to understand and to take care of the typical shipping lanes and what temperature expose the biomaterials so as to encounter variable conditions, keeping touch with the shipping partners to manage appropriate packing, and reduce the delay in logistics process.

In addition, the shipping partners are to be convinced for phase III therapeutic proteins trial materials for proper recording of variable temperature that may be helpful while assuring the quality sample for safe investigation and analysis.

7.3.6 Carrier Logistic

The supply chain network for handling livestock products from the site of production to the retail dealers is relatively smaller as compared to global logistics systems. However, with the development of modern cooling technology, and its application for biologics products distribution under strict Regulatory Acts, third-party logistics are the factors responsible for high cost. But, the cold chain SCM systems have developed confidence among the end users for the safe use of life saving drugs. Currently, in order to supply livestock derived products within the possible shortest time, some of the global air cargos have the provision of good facilities of third-party management systems with instant tracking facilities. Besides this, the logistics providers and freight forwarders have also well network based control towers having well-trained staff to manage with perfect efficacy.

7.3.7 Packaging Technology

Cow and goat milk therapeutic proteins are also belonging from biologics and need the same packing as practiced for biologics. The biopharmaceuticals are highly sensitive to elevate temperature, and loading unloading stresses. So, the manufacturers are having well- trained engineers and technically expert workers to handle active and passive containers for safe transfer of products to different geographic location. The active containers are having well provision of temperature controlling instruments to adjust with variable temperature during transition process. The active containers can be reused for time and again, even under adverse conditions caused by flight and trucking scheduled. The passive cryogenic containers are filled with specific chilling materials like dry ice or compressed liquid nitrogen with release devices. The passive containers are one time use system, and cannot withstand for a long time, in case of any change in flight or trucking scheduled. The overall cost of a shipment is a net balance between the cost of the containers, the cost to ship the containers, and return logistics or disposal costs.

The manufacturer, mainly dealing with therapeutic proteins and dry milk powered with lactose free, should pay special attention to the maintenance of containers and governance involved in operating SCM in a most efficient manner. Such practice can also improve product quality, enhance regulatory compliance, and contribute to the overall state of control by preserving biologics stability, ensuring sterilization and customizing delivery systems for market differentiation.

Thus, enhanced regulations and new guidance on current good manufacturing practices (cGMP) are also driving drug and packaging manufacturers to improve production techniques. The therapeutic proteins are often more sensitive to interactions with vial and syringe components, which can induce protein aggregation or denaturation. Therefore, qualification and control of materials that hold bio-therapeutics are critical to development and manufacturing. For highly sensitive biotech drugs, a packaging solution needs to offer protection from contamination and ensure consistency in every aspect that could affect the long-term drug stability.

Reducing variation in drug packaging can deliver measurable economic value by minimizing the risk of leachable in the drug solution. Packaging components come with certified extractable profiles, which provide proof of each component's biochemical consistency at a new level of control. Packaging manufacturers can also identify and control sources of potentially contaminating particles through inspection programmes and continuous improvement efforts. Thus, this will be helpful in reducing or eliminating the overall expense and exposure to regulatory risk associated with component preparation, sterilization, and end-of-line rejections. Industry leading suppliers can guide drug product manufacturers through such issues and their potential implications.

7.3.8 Cloud Computing

The efficacy of supply chain management systems including the value chain operation protocol has enormously increased due to the advancement in modern information technology (IT). The basic IT is even applicable to biologics SCM system in order to provide the products to wholesale dealers or retail dealers or end user strictly on time bound scheduled with safety guarantee of desirable products. The IT sector has become a valuable boon to the industries for their successful goods transition to different locality, timely. The IT sector works with mutual and high level collaboration with organization or manufacturers for timely discharge of products. For this, the IT sector must be compatible with business aims and objective for better upgrade dynamic business transition.

The collaboration in SCM is inevitable, and strategic alliance in resourcing can help biopharmaceutical manufacturer to achieve a sustainable development pattern under proper management and monitoring process. Collaborative investment and attitude have helped firms differentiate from competitors by enhancing their relationships with suppliers and customers. The main target for collaboration is to how efficiently and creatively connect resources and expertise with alternative and integrated development model. Manufacturers having positive attitude for integration their resources with IT can provide timely, update information with key stakeholders in improving supply chain management process.

Since 2007, use of cloud computing tools and techniques has started gaining popularity due to its ability to provide services to geographically separated clients on demand basis. This tool can be used for supply chain management of variety of materials including biopharmaceuticals. Fortunately, cloud computing is now offering service to small industries same as to larger companies. Cloud computing characteristics of low cost and easy accessibility have the potential of developing collaborative relationship among members of the supply chain because partners can easily implement cloud based applications instead of purchase and install expensive software allowing organizations to work together faster.

7.4 Supply Chain Risk Management

Livestock (goat and cow) milk therapeutic proteins belong to biologic class of drug and their safety management is as per the practices for biologics. Due to the sensitivity to higher temperature, it is necessary to transfer these biomolecules through temperature logistic system. Thus risk management is an inevitable factor while operating logistic system for safe distribution of biological products to different location. A manufacturer needs hurdle free administration while transferring the products from site of production to discharge points present at different locality under the regulatory guidelines, well-managed QC and QA, and well-reputed organization culture. Risk management is a complementary factor closely associated with SCM and value chain. Risk management practice is being carried out by well-trained security personnel having deep knowledge about logistic operation system. So, it is necessary to have close association with risk management activities which can be embedded and maintained throughout the organization. Meticulous planning, performance process, system management, controlling, reviewing, and reporting are closely linked factors with risk management. Generally, risk management awareness develops with intense practical and experience.

Product safety, security, consistency, and quality are the major challenges to SCM and logistics operation. So, risk management plays a critical role while transferring the products from site of manufacturing to retail dealers. Both the USA and Europe follow ICH Q9 guidelines for SCM and logistics processes. Microbial contamination is a major risk factor in biopharmaceutical production industry. Mostly, the biopharmaceuticals are contaminated during hold times, either at processing stage or as inventory in storing place. The primary factors that affect the risk of microbial contamination during hold times are due to the growth promoting properties of the in-process materials, initial bio-burden level, and storage condition. It has been noticed that microbial contamination is the most prevailing factor for the failure of risk management. Available risk scoring tools such as failure mode and effects analysis (FMEA) should be properly managed with suitable modification, on prevailing factors. Continuous supply of products components on quality base for manufacturing process is a key factor to avoid risk in manufacturing process. So, raw material control strategies play a crucial role. Selection of honest persons, solvency, duration of business, and vendors' in-house quality system are important issues to be taken care while negotiating business strategies with vendors. The vendors should have their business continuity plans (BCPs), as per international standard.

7.5 Logistics Operation

Livestock proteins, milk powder, and therapeutic by-products are highly temperature sensitive and need utmost care during transport process. The logistic system is managed under strict regulatory guidelines to make it available to wholesale dealers or retail dealer to meet the local market demand.

Promoters adopt efficient business techniques and skills to improve business performance at global market. Thus, supply chain is the entire operation process such as network sourcing, supply pipeline management, value chain management, and value stream management [56–58]. The logistics systems management is a part of the entire supply chain management operation complex. So it is regarded as a crucial factor for the bio-pharmaceutical manufacturer to gain competitive edge.

In true sense, logistic system operation group is a separate entity being responsible for distribution of products received from the manufacturer under perfect regulatory acts, and as instructed by the experts from the companies. In fact, the concept of logistics management of perishable bio-pharmaceuticals has received attention, recently. Logistics is the management of the flow of goods between site of manufacturing and the end user in order to fulfil the requirements. The logistics of temperature-sensitive livestock products usually involves the integration of information flow, material handling, production, packaging, inventory, transportation, warehousing, and often security. The complexity of logistics can be modelled, analysed, visualized, and optimized by dedicated simulation software. The minimization of the use of resources is a common motivation in logistics for import and export.

7.6 Order Management

Currently, many multinational groups like Dairy Goat Co-operative *(DGC), Ausnutria Dairy, Baiyue Youlishi, YaTai-Precious, Red Star, Guan Shan, MilkGoat, Herds, Fineboon, Jinniu, Shengfei, ShengTang, Holle, FIT, Vitagermine, Xi Yang Yang, Baiyue Dairy Group have started following Real-Time Value Network (RTVN) for increasing the efficacy of order management process for timely delivering the products. In RTVN multi-parties belonging from different companies work together with mutual understanding and co-operation. In this process both the purchase order and sales order are synchronized to meet the timely demand of market located at different place.* For this process integrated network service is maintained between buyers, sellers, movers, shippers, and service providers.

In single order system the supply of products may take a day to week, even more than weeks and months, but RTVT integrated business plan for every item at every location is timely managed. In RTVT system the multiple orders are automatically optimized and executed with time bound programme. The cloud computing network in real time helps the company and trade partners on demand based SCM operation system, efficiently. Additionally, real-time value network is also helpful for storing of inventories, and supply chain process on order basis, and also monitoring a stock-keeping unit (SKU) located at different location.

The Real-Time Value Network currently is composed of over 30,000 trading partners. Organizations that connect to the network gain the potential ability to transact with any of the organization already on the network, offering a true many-to-many global network.

7.6.1 Inventory Planning and Control

In general, inventories are stored goods having economic value, and are safely held in different purposes for the dispose of products or may be utilized for the maintenance of factory as consumables spares, or work in progress and finished goods. The livestock farming or industry inventories are quite different from the other industries from reutilization point of view. The storing provision of inventories is arranged near by a business location of a manufacturing unit for its easy accessibility on need base. Inventories are variable in nature. For example, a local dealer keeps the items on the basis of local demand, but it is different in case of a manufacturer who keeps inventories for factory maintenance and running. In addition, it must maintain some supply of finished products to meet the demand of public, as and when required. For effective and sustainable operation of manufacturing unit the inventories are grouped as (i) raw material, (ii) work-in process, (iii) finished goods, and (vi) MRO goods.

(i) Raw Material

The raw materials required as inventories for livestock therapeutic proteins manufacturing industry are mainly for upstream process, downstream process (for quality up gradation), and for polishing and packing. The upstream process inventories of a livestock therapeutic protein and milk powdering units need different parts of stainless steel reactors and its accessories like filter sets, high pressure regulatory pneumatic valves, etc. for immediate maintenance, and keep the system in nonstop operating condition, as the host cells are highly temperature and oxygen sensitive. However, animal based bioreactor, especially goat as biological bioreactor is different from the stainless steel bioreactor. Extra care has to be taken to nurture the transgenic goats in specially designed farmhouse where the breeding purity is maintained by well-trained animal genetic scientists. The downstream process for the therapeutic proteins isolation and polishing is a conventional downstream process, as practiced in a biopharmaceutical industry engaged in biologics manufacturing process. Some of the steps in downstream process also need nonstop operation, and for this, keeping accessory as emergency inventories is a primary practice in downstream operating process.

The inventory items may be commodities of biological origin that the farm or its subsidiary has produced or extracted. There may be some objects or elements that the firm has purchased from outside the organization. Even if the item is partially assembled or is considered a finished good to the supplier, the purchaser may classify it as a raw material for manufacturing process. Typically, raw materials for cattle and goat farming are commodities such as farmhouse construction materials, source for grazing, green feeds and calculated feeds, milking machines and other accessories required for nurture livestock.

At corporate level Material Requirement Planning (MRP) is the beginning stage of manufacturing process. MRP is a planning scheduled for the requirement of variety of inventories to go ahead with Manufacturing Resource Planning (MRPII) to develop entire protocol for sequential manufacturing process for a new product development. Primarily, it is meant for operational planning and financial planning in units. Currently, the goat milk therapeutic proteins manufacturers are having the practice of MRP and MRP II in order to maintain the quality and quantity of milk products, both from normal animal and genetically modified goats for milk therapeutic proteins.

(ii) Work-In-Process (WIP)

In cow and goat milk processing unit the biological derived materials or the biomaterial under process for products such as milk powder, therapeutic proteins development are known as work-in-process. In other words, cow and goat derived products are considered to be work-in-process. WIP also includes labour and overhead costs incurred for products that are at various stages of the production process. WIP is a component of the inventory asset account on the balance sheet, and these costs are transferred to the finished goods and eventually to cost of sale.

(iii) Finished Goods

Non-transgenic cow and goat milk product (mainly instant lactose free dry baby milk) and transgenic goats milk therapeutic proteins are manufactured under international guidelines of regulatory acts, and packed and disposed through SCM or third-party under fully temperature controlled logistic system.

These goods have been inspected by QC & QA experts before transport or shipment. This practice has been operated to maintain consistency and safety of products in order to develop confidence among end user at global front on manufacturer. Inventories can be further classified according to the purpose they serve. These types include transit inventory, buffer inventory, anticipation inventory, decoupling inventory, cycle inventory, and MRO goods inventory. Some of these also are known by other names, such as speculative inventory, safety inventory, and seasonal inventory.

(iv) MRO Goods

Equipment and machineries and their respective parts used for maintenance, repair and operation are under MRO inventory category items. The MRO inventory items are utilized in the manufacturing process, but are not part of the finished goods being produced. For an example, the lactose free milk powder and therapeutic proteins from transgenic goats and sheep are of short life span. So, the equipment and machineries (parts of fermenter, milk spray drying parts and any other instrument related to milk processing), used for this process are to be kept in stock in order to meet the emergency need, if any damage occurs to the original machineries or their parts in operation condition.

7.6.1.1 Transit Inventory

Any cattle and goats products in the process of transferring from site of production (farmhouse or milk processing unit) to warehouse after proper packing, and subsequently trucking or shipping are known as transit inventories or pipeline inventories. Merchandise shipped through truck or rail may take a lot of time for shifting from regional warehouse to wholesale dealers or retail dealers. As the goat milk products are prone to elevated temperature the delay caused due to such transfer process may cause for losing the property of quality before reaching to end users. Thus flight based logistic system is safe and makes the availability of goods to whole sellers timely.

7.6.1.2 Buffer Inventory

Under certain crisis period cow and goat milk processing unit for therapeutic proteins manufacturing is not able to meet the demand of wholesale dealers or retail dealers, timely. In order to overcome with such crisis the manufacturer has to keep sufficient stock for manufacturing process. Safety stock or buffer inventory is any amount held on hand that is over and above that currently needed to meet demand. Generally, the higher the level of buffer inventory, the better the firm's customer service. This occurs because the firm suffers fewer "stock-outs" (when a customer's order cannot be immediately filled from the existing inventory) and has less need to backorder the item, make the customer wait until the next order cycle, or even worse, cause the customer to leave empty-handed to find another supplier.

7.6.1.3 Anticipation Inventory

Sometimes, cow and goat milk processing industries go for purchasing sufficient feedstock to face the crisis like price increase, a seasonal increase in demand, or even an impending labour strike. This practice is common with retail dealer, who routinely build up inventory months before the demand for their products will be unusually high. Generally, livestock housing should have sufficient quantity of anticipation inventory when demand is low, and keep the workers busy during slack times. The anticipation inventories help the promoter when demand pick up in the market. Thus, the firm does not take botheration to manage market demand without putting the labourers to do overtime work. It is also helpful in avoiding layoff costs associated with production cut-backs, or worse, the idling or shutting down of facilities. This process is sometimes called "smoothing" because it smoothes the peaks and valleys in demand, allowing the firm to maintain a constant level of output and a stable workforce.

7.6.1.4 Decoupling Inventory

Manufacturing of the therapeutic proteins from transgenic cow and goat milk has been gaining popularity among biopharmaceutical manufacturers. The goat is used as bioreactor for expressing biologics through mammary gland. In this process the complex upstream process (maintainers of host cells and expensive stainless steel bioreactor) is bypassed. However, the downstream process for obtaining target biologic(s) is as complex as practiced in separation and purification of proteins associated with conventional fermentation process. In order to reduce the cost and process time the various stages of downstream process are linked in partially continuous manner.

However, in practice, a promoter wants to keep the continuity and consistency in manufacturing process. The work-in-process could be manageable with the proper arrangement of a decoupling inventory that serves as a shock absorber, the system against production irregularities. As such it "decouples" or disengages the plant's dependence upon the sequential requirements of the system (i.e., one machine feeds parts to the next machine).

The more inventories a firm carries as a decoupling inventory between the various stages in its manufacturing system (or even distribution system), the less coordination is needed to keep the system running smoothly.

7.6.1.5 Cycle Inventory

Generally, livestock farmer's intension is to sell goods irrespective of selling them to retailers or find out some ways and means to manage stock inventory balance without any financial burden. The cycle stock inventory is most crucial part of overall inventory since it is the first choice of customer purchases.

Cycle stock inventory is a part of an inventory that helps in perfect monitoring the seller cycle with safe maintenance of sales order with readily available inventory present in store. In due time cycle stock inventory is replaced by new items in order to maintain a reasonable stock for smooth sale. The inventory cycle and GDP are complementary with each other. Any variation of GDP causes accumulation of selling of stock. It is a regular practice of a promoter to monitor balance inventory holding or carrying costs with the cost incurred from ordering or setting up machinery which is known as economic order quality (EOQ).

Cycle inventories are also called as lot-size inventories. It is also referred to the quantity of an item ordered for timely delivery with safe and security. It is also necessary to take care of quantity price discounts, shipping cost, and setup cost to market it more economical to sell.

7.7 Concept and Significance of Inventory Management

The role of inventories management is most crucial in successful performance of business transition and timely supply of products to wholesale dealers or retail dealers. The milk therapeutic proteins and lactose free dry baby milk powder are highly temperature sensitive. Thus, a dairy manufacturing industry and transgenic goat milk processing unit need critical attention from inventories points of view. Generally, these types of industries are having highly integrated manufacturing process to minimize the cost of production. Some key aspects like SCM, value chain monitoring, cold chain logistic systems with the well provision of third-party logistics groups and well-trained inventories management manpower form the backbone of sustainable business deliver function and part and parcel of marketing management as well as finance controllers.

7.7.1 Warehouse Management

Warehouse management is a managerial art of system monitoring techniques rather than a commercial orient. The elevated temperature sensitive transgenic goat milk derived therapeutic proteins are critical to manage through SC or value chain systems. In addition temperature control regional warehouse is needed, till the products are properly transferring for shipment through third-party logistics systems.

Warehouse management is a key practice for delivering the cow and goat derived products with safe conditions. The warehouse should be fully temperature controlled with the provision of digital controlling system to keep the products on manufacturer's need base. As the biologics are lifesaving drug, small variation of temperature may spoil the quality of the goods. That is why large companies cannot operate without state-of-the-art warehouse management solutions.

7.7.2 Transportation vs. Supply Chain Management

Transportation refers to the movement of product(s) from the site of manufacturing unit to the wholesale dealers located at different places. Transportation is a part of supply chain management which includes procurement of goods from the site of manufacturing, storing in well-managed warehouse before shipping to wholesale dealers present at different location. Currently, with the emergence of real-time data networks, radio frequency identification technology (RFID) tags provide accurate status reports and point–point inefficiencies in the goods supply chain. Supply chain management (SCM) is the broad range of activities required to execute and monitor products movement from the site of production to the wholesale dealers in a cost-effective process with the safety of goods in both quality and quantity wise. For SCM, a good manufacturer has its own team, well trained in handling the goods for shipping.

The transgenic cow and goat milk therapeutic proteins come under biopharmaceutical category which also need safe and maximum secure during supply chain managing process. A manufacturer wants guarantee on safe and security of goods till it reaches the end users. Thus, it is of primary importance to protect and validate goat's therapeutic proteins from the forged product market. Overcoming such forged market use of RFID technology is the best solution to keep track with the instant status of goods during movement in supply chain process. In addition, RFID can also be useful in information hiding techniques to get information on completion of expiry date, tamper detection, and fraud detection and further prevention.

7.7.3 Flow of Manufacturing Costs

Financial transition of investment and its flow manufacturing cost is defined as the chargeable cost of products starting from the procurement of raw materials incurred for manufacturing to the finished products. This process is applicable to all type of business management practice, including livestock and livestock derived product supply chain management system. So, the basic principle and application of the supply chain management in livestock and livestock products selling process is of very significance. When the livestock are grown for the production of animal protein, dry milk powder and other by-products, raw materials or feedstock are stored and passed on to farmhouse, the cost then moves to work-in-process inventory in order to meet the requirement of the cost of labour, machining, and overhead cost, and these costs are added to the base cost of the produce/feedstock. After the production process, the costs move to the finished goods inventory classification, where the final quality product(s) are stored prior to sale through suitable logistic systems. In modern livestock farm houses, course of process flow, the costs are computerized at the beginning, and additional costs procured in value chain process are added to calculate overall cost of the product while preparing the final balanced sheet. The flow of manufacturing cost and financial profile mainly depend on the inventories associated with supply chain management process. Inventory is where the manufacturer's measure of its funds invested. Inventory typically consists of finished products for sale, raw materials under process for producing finished product. To calculate the profit a company gain, it must track sales revenue as well as all of the costs involved in procuring raw material or livestock. Accordingly, the manufacturer profits consist of the money remaining from sales after the company has covered all its costs, including buying of its inventory.

In flow of manufacturing cost various terminologies are being used to define the flow of manufacturing cost at different steps of goods transition from the site of production to main store warehouse, and subsequently to wholesale dealer. Two most common terminologies, i.e. FIFO (first in first out) and LIFO (last in first out) are in use in main goods transition.

Generally, the immediate manufactured goods are recorded as inventory. The costs of inventory items immediately manufactured are changed to expense when the same are sold last. FIFO, on the other hand, assumes that the first items put on the shelf are the first items sold, so the old stocks are sold first. This system is generally used by the bio-pharmaceutical manufacturers where the inventories are perishable or prone to long storing practice. It is mainly preferred accounting method when the cost of product is rising in the market. Mainly the cost flow depends on the dynamic nature of manufacturing process, work in progress, finished goods inventory, and quantitative procurement of raw materials or feedstock for manufacturing process. LIFO and FIFO have a different effect on a manufacturer financial condition.

A bio-pharmaceutical company, like other manufacturers, uses the FIFO system to trace its inventory by reporting LIFO method while preparing final balance sheet. By this the gap between FIFO and LIFO can be bridged by using LIFO as reserve account. In case of price rising, credit balance in the LIFO reserve can be created causing reduction in inventory cost. Mostly, the manufacturers or business people use FIFO, or Standard Cost Method, for internal use, and LIFO method for external reporting, as the case tax preparation. By reporting in LIFO method, this helps in reducing the tax at the time of price rising. Storing of goods in anticipation of demand and their marketing at proper time are depending on the efficacy of supply chain management system. In FIFO system, the manufacturer tries to clear maximum amount of goods, and the products left without sale are mixed with the subsequent lot assuming that its quality matches with the existing leftover batch having sufficient expiry date.

The LIFO liquidation can be computerized in different ways. Generally, a manufacturer follows the LIFO liquidation by using the last in first out (LIFO) of inventory costing, and then liquidation its older LIFO Inventory. In case of current sale is higher than the manufacturing cost (including the cost of raw materials), the promoter should liquidate the stock inventory. This practice will be helpful in gaining financial profit in net operating income. The biopharmaceutical industries, generally, follow LIFO methods for immediate disposal of products in the market. This method is followed when the period of rising prices or period of inflation, when the cost to purchase inventory increases over time. The tax burden of a pharmaceutical company is reduced when the higher priced inventory matches to the current revenue, increasing cost of goods sold (COGS) and decreasing earnings before (EBT), LIFO method is used for asset management and valuation for assert management while acquiring selling or disposing the inventory goods. If an asset is sold for less than it is acquired, then the difference is considered a capital loss. If an asset is sold for more than it is acquired, then the difference is considered a capital gain. Using the LIFO method to evaluate and manage inventory can be tax advantageous, but it may also increase tax liability.

7.7.4 Trace Link and Serialization of Animal Proteins

Top multinationals such as biopharmaceutical, pharmaceuticals, and biotechnology industries and dairy industries implement traceability and serialization capabilities to meet regulatory requirements, quality, and recall purposes. It requires a community that can share that information effectively and collaborate with law enforcement at national, state, and local levels. After the implementation of traceability and serialization practice US pharmaceutical industry has achieved over 50X reduction in pharmaceutical theft over the years. In this connection theft incident occurred in an Eli Lily warehouse ($76 m of drugs stolen) is most alarming for pharmaceutical and biotechnology based industries [52].

The primary objective of traceability is typically not theft control, but rather quality control and enabling recall capabilities. The packaging of biopharmaceutical products should carry tracking microchip to understand the condition of goods at instant time. From tracking hologram one can know the instant quality of goods, information on expiration date, batch or lot number, and recommended storage condition, direction for handling and precaution that may be necessary, detail about the manufacturer, and recommended storage conditions.

7.7.5 Expiration Date

In transgenic goat milk production and therapeutic proteins isolation regulatory acts have to be followed, in order to maintain safety and quality of the products. The packing should carry a sticker mentioning the date of expiration. An expiration date means the maximum life of a biological origin product, after which it is strictly advised for no further use, and should be destroyed, without creating any environmental problem. Arbitrary expiration dates are also commonly applied by companies to product coupons, promotional offers, and credit cards. In these contexts, the expiration date is chosen for business reasons or to provide some security function rather than any product safety concern. Expiration date is often abbreviated EXP. The legal definition and usage of the term expiration date is varying between country and nature of the products. The expiration date for each biological origin product should be computed from the date of the initiation of the potency test. Prior to licence, stability of each fraction should be determined by methods acceptable to Animal and Plant Health Inspection Service.

7.8 Impact of New Technology Development on Supply Chain

Steady increase in demand of goat milk derived dry baby milk powder (lactose free) and milk derived therapeutic proteins has pursued the global multinational groups, the efficacy of increasing manufacturing process and supply chain management in order to meet the demand of end users without much financial burden. Linking the supply chain with information technology (IT) not only increases the efficacy of goods transition from the site of production to the site of use, but also maintains the quality of the products and timely delivery consignment [59]. In reality, IT could be able to create a link with supply chain and help in keeping close touch of the consumer with manufacture process and quality control information. In addition, the alignment of IT goals and objectives with strategic SCM can increase efficiency, productivity, and profitability [60]. By integrating SC with IT immense benefit would achieve in product development, procurement, manufacturing, physical distribution, customer relationship management, and performance measurement [61]. This sort of practice supports the promoter by providing the means to link technology and people and trying to align the technology with the capabilities of the organization and among its trading partners [62].

Besides getting quality product, the customers also like to interact with supplier to develop confidence in consistency in supply chain process, and competitive status of the product as compared to the other manufacturers. A good competitor should have flexibility in business dealing, and negotiation ability in marketing guarantee of cost, quality, and timely delivery. This could be only practicable through advanced information technology [63]. Thus supply chain information systems have key role in speeding up the business transition and improving decision making and productivity in most efficient way with controlled budget [59].

By linking IT a company can integrate internal business function and keep track with the activities of company well in time. This can help the promoter to fulfil the requirement of customer and catalyse the quality control process [64]. SCM systems with the help of IT help logistics management, transportation management, strategic planning, warehousing, inventory, manufacturing, supplier management, and customer management [65]. Enterprise Resource Planning (ERP) systems are included as part of the broader SCM software. ERP systems are employed to integrate business processes, by organizing, codifying, and standardizing business processes and data [66]. Functionally ERP helps in systematic way of data management and transferring data within internal system by preventing the expense on transportation of data from one department to another. ERP reports can be used to forecast production and make decisions. Another key supply chain process is customer relationship management (CRM), which is the management of relationships between the organization and its customers.

A critical issue for internal integration of business processes is the integration of ERP with CRM. Recently, many ERP vendors provide an ERP-CRM integration package, since many enterprises have expressed an interest in integrating a new CRM system with their legacy ERP system. Moreover, the establishment of ERP and CRM systems should be a primary concern of an enterprise that is interested in embarking on e-business [67]. Generally, company shares critical information on demand prediction, actual orders, and inventory level in an efficient manner [68], while protecting each company's proprietary data. The adoption of inter-organizational information systems (IOS) for supply chain management has been proposed to enable external integration [69, 70]. The most common form of IOS is electronic data interchange, which permits instantaneous computer-to-computer transfer of information. There are some technologies that, overall, are considered of high relevance for future supply chains, such as cloud solutions and mobile devices.

Even though security concerns are under consideration yet again, cloud computing has become an integral part of today's supply chain management solutions portfolio and remains of interest. The main benefit is that they address companies' investment capability and functional requirements at various stages of their development and apply to a number of business areas in global logistics. Mobile computing and mobile devices have gained greater importance in recent years due to their ability to efficiently support supply chain processes "on the go". Their technical advancement has shown significant development in their scope of capabilities, but their most valuable feature continues to be the ability to integrate into ERP and operational systems to gather and share data in a variety of forms and formats.

Comprehensive, sophisticated supply chain management systems meet the requirements of supply chains in the twenty-first century: adaptability, collaborative ability, transparency, compliance, and speed in all processes. Streamlining processes and cutting costs in the long run are all vital for future business growth. By choosing the right software solution, supply chain managers can future-proof their business and gain a competitive advantage.

7.9 Green Logistics and Supply Chain (GLSC)

Consciousness on conservation of blue planet is the first priority associated with each and every activity. Activity leads to industrialization for products development process for the services of humanity. But, the question on environment stability and sustainability is a complementary factor closely linked with manufacturing process. The various issues related to environment are directly resulted from the mismanagement of industrial activity, starting from the raw material procuring, manufacturing, and disposal of goods through logistic system. During the process of goods transition, the chances of spoiling of goods, damage of shipping system on its way to destination, and microbial contamination of packing materials are some of the major problems associated with logistic system operation. The introduction of digital based logistic systems has significantly lessened the burden of environmental pollution. Logistics systems development and operation without any wastage generation is known as Green Logistics and Supply Chain (GLSC). Green supply chain is the supply chain management with similar objectives and core implications. Green logistics as well as supply chain management is also usually referred to as "sustainable" management.

In biopharmaceutical industries, besides generating solid, liquid, and gaseous (green-house gases) biological wastes generation during goods transition is also a major environment issue. So, the manufacturers are keen to incorporate advance technology during product packing to avoid any biological waste generation by leakage of products or due to microbial contamination. It is aimed to reduce environmental pollution and resource consumption arising from logistics activity so as to realize a "win-win" consequence in logistics development and eco-environmental conservation. In this process, traditional practice of supply chain is integrated with environmental conservation so that waste generated during any supply chain can be well recycled in order to induce susceptibility in supply chain management practice and bring profit to manufacturer (Fig. 7.6). Microbial processes activity for the production of biopharmaceutical, health care product like nutraceuticals, food supplement, food additives, antioxidants, etc. may threat to the environment in terms of carbon monoxide emissions, discarded packaging materials, scrapped toxic materials, traffic congestion, and other forms of industrial pollution [71].

Thus, all procedures related to hazardous and non-hazardous wastes should be controlled and monitored in logistic system management process. The main goal of control and monitoring is to anticipate, identify, and verify unintentional release of any biological or non-biological materials. Minimal monitoring should include periodic physical inspections of waste generation sources, storage areas, means of transport, waste related documentation, and proper labelling and control activities establishment.

A typical field in green logistics and supply chain management is reverse logistics, sometimes called closed-loop supply chains, in which the rear reverse flow so fused products back to manufacturers [72]. People were well aware of reutilization of goods, if any left during logistic transition process [73–75]. With

Fig. 7.6 Green supply chain management systems

the advancement of IT development, control strategy and pull control strategy concept, and explore the superior inventory strategies for hybrid manufacturing/remanufacturing systems with a long lead time for manufacturing and a short lead time for remanufacturing were developed to upgrade supply chain system in a more efficient manner [76, 77].

Currently, the overall concept on business philosophy has drastically changed. A good multinational manufacturer not only targets for profit but also wants to keep this habitable world unaffected from stress and strains caused by creating minimum issues that can challenge the environment sustainability. For this, only taking care of waste free manufacturing process would not solve the problem. Follow-up and maintenance of green logistics and supply chain (GLSC) are an integral part of total manufacturing activity, in sustainable form [78].

Hence, green logistics and supply chain (GLSC) impact on environment is minimum and no harm on sustainability performance of supply chain. During the beginning of the 1990s, competition among big multinational increased in order to impress the consumers from aesthetic and social responsibility points of view on product purity maintenance, without adding any issue to environment cleanliness [79, 80]. Having these practices in mind, firms develop environmental management strategies in response to the dynamic nature of environmental requirements and their impacts on supply chain operations [81]. A successful supply chain management is a coordinate attempts, being controlled by multiparty systems include manufacturer, suppliers, and wholesale dealers [82].

7.10 Quality Control

Livestock proteins, milk powder, and therapeutic protein are produced from the living organisms for human use. So, it is necessary that the livestock protein production need quality control while in production stage. Thus, the livestock primary products are susceptible to various stresses and strains, it is necessary to have control over the products quality in various stages of production. Sustainable development of primary product from livestock is mainly dependent on quality control as per the norms prescribed in the international regulatory acts. Quality control is unavoidable while planning for successful business that delivers products and meet the customer's expectation. It also gives additional support and confidence while delivering the goods with minimum waste and operates at high levels of productivity. It is mandatory for manufacturer to have perfect protocol of quality control before starting manufacturing process. The quality control encompasses with two words i.e., quality and control. Quality (mainly as biologic drug) represents the overall character of a product on the basis of its size, shape, durability, efficacy, and stability in formulation. Thus, quality is the backbone of a biopharmaceutical manufacturer which involves in biologic manufacturing process, and well equipped with analytical instruments, and trained technicians to keep the quality as per the norms of international regulatory acts.

7.10.1 Objectives of Quality Control

The main target of quality control is to manufacture consumer acceptability product (s) and make it timely reachable to end user with the guarantee and safety of quality product. Besides, the quality control gives the guarantee of flawless manufacturing process by suggesting improvement in manufacturing process, on the basis of user's demand. Planning the manufacturing protocol by giving maximum emphasis on quality of the product and service offered to the customer's requirement in a consistency manner are the unavoidable practice of a manufacturer for success.

So, most of the multinational companies have super specialized cell with well-trained scientists to have control over production. One of the most popular examples in this regard is the starting business philosophy of Sony Corporation, which is to bring unification in company workers with a farm spirit of team work with complete involvement which could lead the company to achieve unexpected gain and fame in global market. American and European company have to spend tough time to bring quality control over manufacturing process. For maintaining and enhancing quality, the manufacturer has to adapt two techniques, i.e. quality control and quality assurance, and make sure that the end product or the service meets the quality requirements and standards defined for the product or service.

7.10.2 Quality Control Management

Quality control and quality assurance are two separate practices in manufacturing process and are major pillars of success in business. Generally, the manufacturers possess the quality control manual being approved by international regulatory acts. The quality control manual outlines the quality focus and objectives in the organization. On the basis of guidelines, the different department use to coordinate the manufacturing process to keep sustainability in business transition.

An organization has to follow the best suitable method for quality control being advised by highly professional quality control group. Some organizations believe in the concepts of total quality management (TQM) and some other organizations want to have their own internal and external standards. Generally, the procedural processes for managing activities are carried out on the basis of norms as prescribed by International Standard Organization (ISO). So, it is highly necessary for a company to adhere to the quality requirements of ISO.

Since standards have become a symbol for products and service quality, the customers are now keen on buying their product or the service from a certified manufacturer or a service provider. Therefore, complying with standards such as ISO has become a necessity when it comes to attracting the customers. The importance of properly established and managed quality control and quality assurance systems with their integral well-written SOPs and other quality documents for the achievement of company business objectives cannot be ignored. They serve as a passport to success and maintain sustainability in business.

7.11 Quality Assurance

Quality assurance (QA) is the practice of monitoring and preventing the various issues related to product deformation and defect during manufacturing process. In addition, QA keeps in touch with the product movement through supply chain or value chain systems. The ISO defines the QA practice as: “part of quality management focused on providing confidence that quality requirements will be fulfilled”.

The QA is governed by two principles: "Fit for purpose" on the basis of its suitability for the intended purpose, and "right first time" to eliminate mistake. In QA practice intense care is taken before the background of designing and developing a novel technical product.

Often, quality assurance and quality control are used interchangeably to interpret by ensuring the quality of a service or product. The term "assurance" is used for the implementation of inspection and product testing as a measure of quality assurance, depending on the nature of the manufactured process and product development. The term "control" is used to explain the fifth phase of the Define, Measure, Analyse, Improve, and Control (DMAIC) model. DMAIC is a data-driven quality strategy used to improve process.

Quality assurance brings confidence among users in using specific branded product. Sustainability of product quality is to monitor by quality assurance practice. Thus, fully autonomous is given to quality assurance cell without any threat and pressure from higher administration body of the company. The quality assurance should operate under strict guideline of international regulatory acts like GxPs, Current Good Manufacturing Practice (cGMP), GCP, etc., and local, national, regional, and international, legal, ethical, and regulatory requirement. Quality assurance cell has much responsibility to stream line the protocol for quality control on the basis of end users requirement. Mostly, a company prefers to adapt ISO or CMMI for quality assurance. In CMMI practice a company has to define its own internal processes and adhere by them. In highly established multinational company sophisticated software system is used to understand the instant nature of quality assurance while a product is in manufacturing process. It is necessary to keep update information by quality assurance group by regular inspection, while the product is in process, and also moving through SCM.

The stability test is an important criterion for quality assurance practice. By stability test one can understand the change in quality of such product in response to various physical factors like temperature, humidity, light while storing and transporting to destination. The standard protocol for testing is well defined on the basis of effects of climatic conditions in the three regions of the EU, Japan, and the USA.

On the basis of global climatic data, one can easily calculate the instant climatic condition of a specific region. The world can be divided into four specific zones based on the climatic conditions. This guidance addresses climatic zones I and II. The principle has been established that stability information generated in any one of the three regions of the EU, Japan, and the USA would be mutually acceptable to the other two regions, provided the information is consistent with this guidance and the labelling is in accordance with the national/regional requirements.

Stress testing can helpful in understanding the degradation nature of biologic drugs which can be helpful while the biologic drug moves from the site of manufacturing to the wholesale dealers. Generally, the stress testing is carried out on a single batch, selected randomly. The stress test is carried out at different temperature profile (starting from 10 °C increment up to 60 °C), and humidity (e.g. 75% relative humidity) to understand the changeability occurs due to partial oxidation or photolysis. The testing also includes the susceptibility of drug components in response to pH variation, when in solution or suspension. The standard protocol for stress testing is well described in ICH QIB Photo Stability Testing of New Drug substances and products. As per international regulatory acts there should be three minimum testing samples, taken randomly. The capacity of each batch should not be less than the size of pilot level production followed by the same manufacturing process as carried out in turnkey level production.

The testing material for stability study should match with the quality of materials taken for scaling up manufacturing process. Special care is to be taken while using biologic drug samples stored in warehouse or in transport process. The testing should cover, as appropriate, the physical, chemical, biological, and microbiological attributes. Validated stability-indicating analytical procedures should be applied. Whether and to what extent replication should be performed should depend on the results from validation studies.

For long term studies, frequency of testing should be sufficient to establish the stability profile of the drug substance. For drug substances with a proposed retest period of at least 12 months, the frequency of testing at the long-term storage condition should normally be every 3 months over the first year, every 6 months over the second year, and annually thereafter through the proposed retest period. Well-defined protocols are available on storage of biological/biotechnological/pharmaceutical products under variable physical conditions [83].

Generally, the biologic drugs are in conjugation with a second moiety or binding to an adjuvant. So, it is necessary to test the intact structural profile or to study the dissociation nature of biologic drugs at different state of movement in supply chain system.

7.11.1 Legal Basis for Approval of Biologics from Livestock

The legal approval of biologic drugs varies from country to country. In the USA, mainly two statutory bodies, i.e. the Federal Food, Drug, and Cosmetic Act (21 U.S. C. § 301 et seq) (FFDCA) and the Public Health Services Act (42 U.S.C. § 262) (PHSA) govern the drugs approval process. The FDA (U.S) keeps the safety of biologic drugs in order to protect public health. The FDA administrates the FFDCA and PHSA (among other statues). FFDCA applies to all drugs and medical devices, and PHSA applies to "biological products". Marketing approval under the FFDCA is by means of a New Drug Application (NDA) while approval under the PHSA is by means of a Biologics License Application (BLA). The biologics are subject to Investigational New Drug Application (INDA) regulations. Pre-clinical research on new compounds is carried out in a laboratory, using a wide variety of techniques. Promising candidates are then studied in animals, and, subsequently, various clinical studies in humans are carried out following strict guidelines:

Phase I: A small number of healthy volunteers are given the compound to determine mainly that the drug is safe for human use.

Phase II: A small number of patients are given the medicine to assess its efficacy and safety and to ensure that there are no unacceptable side effects.

Phase III: A large number of patients, usually thousands, take the medicine under supervision over a defined period of time, with the results used to establish efficacy.

If the results show the drug to be efficacious and safe, the data are presented to the FDA. The FDA reviews the data, and if the data is acceptable, a marketing authorization is issued. Alternatively, the FDA may request additional studies or reject the application. Following the grant of marketing authorization, the drug product is studied in large numbers of patients in hospitals and clinics to further assess its clinical effectiveness. This stage is called Phase IV or post-marketing study. Safety Assessment of Marketed Medicines (SAMM) studies help identify any unforeseen side effects. In order to be marketed, a biologic requires only proper labelling and an approved BLA that indicates the product has been determined safe, pure, and potent, and that the manufacturing facilities meet the requirements to ensure safety, purity, and potency.

Though biologics have traditionally been subject to much more scrutiny in manufacturing than drugs, those differences are being eroded. Biologics have been approved under FFDCA and PHSA, thus, both NDA and BLA applications have been submitted for biologics. The exceptions are glucagon and follistim that were approved under §505(b) (2), and insulin, which was approved under its own statute for a time. The default approval pathway for biologics now is a BLA, unless the product is a hormone, in which case §505(b) is used.

7.12 Quality Control and Supply Chain Integration

The positive impact of total quality management (TQM) on acquiring successful business in biotechnological and biopharmaceutical business has been observed [84, 85]. It plays a vital role in the development of management practices [86, 87]. It has been noticed that TQM is immensely helpful in upgrading the business transition with fulfilling the complete requirement of customers [88]. In addition TQM increases sustainable competitiveness in business at national and global market [89] and above all as a source of enhancing organizational performance through continuous improvement in organization's activities [90–92]. Henceforth it is strongly recommended to incorporate both the QC and QA practices and procedures in SCM. Thus, quality management is often related with a model of organizational change [93], the implementation of which largely relies on the organization's ability to adapt itself to these principles.

Quality is an important factor when it comes to any product or service. With the high market competition, quality has become the market differentiator for almost all products and services. So, the practice QC has been an integral part of supply chain in production and distribution of product at consumer level. It helps in developing confidence in customer on product acceptability. Without quality control, waste becomes prevalent beyond a tolerable amount.

The microbes contamination is a common problem with biologic drugs, if not properly handled while packing and transport. In order to maintain the quality of product vendors, the materials they provide are often audited by supply chain staff members to ensure raw materials meet specifications. By controlling the quality of production inputs, supply chain managers are protecting the integrity of their company's operations.

Genetically engineered host cells are in use to produce biologic drugs. Quality control helps in screening those genetically modified host cells from the possibility of possessing toxin compounds harmful for human use. The US Department of Transportation prescribes important rules for the transport of hazardous substances. Non-compliance can lead to penalties or fines, which makes QA and QC imperative. The more efficiently and effectively toxic materials are handled in the supply chain, the better for all internal and external stakeholders.

References

1. Alexander FL (1977) Cooperatives and the poor. A development study prepared for the International Cooperative Alliance and the Canadian International Development Agency, the co-operative was initially referred to as Anand Milk Federation Union Limited, hence the name AMUL
2. Tehrany E, Sonneveld K (2010) Packaging and the shelf life of milk powders. In: Food packaging and shelf life, a practical guide. CRC Press, Boca Raton, pp 127–141
3. United States Dairy Export Council (USDEC) (2005) Reference manual for US milk powders
4. Dhanda SJ, Taylor GD, Murray JP et al (2003) Goat meat production: Present Status and Future Possibilities. Asian Australas J Anim Sci 16:1842–1852
5. Gordon K, Lee E, Vitale JA et al (1987) Production of human tissue plasminogen activator in transgenic mouse milk. Bio/Technology 5:1183–1187
6. Meade H, Gates L, Lacy E et al (1990) Bovine aS1-casein gene sequences direct expression of active human urokinase in mouse milk. Bio/Technology 8:443–446
7. Simons JP, McClenaghan M, Clark AJ (1987) Alteration of the quality of milk by expression of sheep b-lactoglobulin in transgenic mice. Nature 328:530–532
8. Hammer RE, Purse VG, Rexroad CE et al (1985) Production of transgenic rabbits, sheep and pigs by miroinjection. Nature 315(6021):680–683
9. Bösze Z, Hiripi L (2012) Recombinant protein expression in milk of livestock species. Methods Mol Biol 824:629–641
10. Yang H, Li QW, Han ZS et al (2009) Recombinant human antithrombin expressed in the milk of non-transgenic goats exhibits high efficiency on rat DIC model. J Thromb Thrombolysis 4:449–457
11. Han Z, Wu S, Li Q et al (2009) Efficient human growth hormone gene expression in the milk of non-transgenic goats. Folia Biol 55:17–22
12. Toledo JR, Sánchez O, Seguí RM (2006) High expression level of recombinant human erythropoietin in the milk of non-transgenic goats. J Biotechnol 17:225–235
13. Han ZS, Li QW, Zhang ZY et al (2007) High-level expression of human lactoferrin in the milk of goats by using replication-defective adenoviral vectors. Protein Expr Purif 53:225–231
14. Toledo JR, Sánchez O, Montesino R et al (2008) Highly protective E2-CSFV vaccine candidate produced in the mammary gland of adenoviral transducer goats. J Biotechnol 133:370–376
15. Sánchez O, Toledo JR, Rodríguez MP et al (2004) Adenoviral vector mediates high expression levels of human growth hormone in the milk of mice and goats. J Biotechnol 114:89–97

16. Xiao B, Li Q, Feng B et al (2009) Expression of recombinant human nerve growth factor beta in milk of goats by recombinant replication-defective adenovirus. Biotechnol Appl Biochem 157:357–366
17. Fan W, Plaut K, Bramley AJ et al (2004) Persistency of adenoviral-mediated lysostaphin expression in goat mammary glands. J Dairy Sci 87:602–608
18. Fan W, Plaut K, Bramley AJ et al (2002) Adenoviral-mediated transfer of a lysostaphin gene into the goat mammary gland. J Dairy Sci 85:1709–1716
19. Archer JS, Kennan WS, Gould MN et al (1994) Human growth hormone (hGH) secretion in milk of goats after direct transfer of the hGH gene into the mammary gland by using replication-defective retrovirus vectors. Proc Natl Acad Sci U S A 91:6840–6844
20. Zhang K, Wang H, Bao Y et al (1997) Human clotting factor IX (hF IX) secretion in goat milk after direct transfer with hF IX minigene into mammary gland by using retroviral vectors. Chin Sci Bull 42:1308–1313
21. Parker MH, Birck-Wilson E, Allard G et al (2004) Purification and characterization of a recombinant version of human alpha-fetoprotein expressed in the milk of transgenic goats. Protein Expr Purif 38:177–183
22. Behboodi E, Ayres SL, Memili E et al (2005) Health and reproductive profiles of malaria antigen-producing transgenic goats derived by somatic cell nuclear transfer. Cloning Stem Cells 7:107–118
23. Schmidt C (2006) Belated approval of first recombinant protein from animal. Nat Biotechnol 24:877
24. Kling J (2009) First US approval for a transgenic animal drug. Nat Biotechnol 27:302–303
25. Edmunds T, Van Patten SM, Pollock J et al (1998) Transgenically produced human antithrombin: structural and functional comparison to human plasma-derived anti-thrombin. Blood 91:4561–4571
26. Ebert KM, Selgrath JP, Di Tullio P et al (1991) Transgenic production of a variant of human tissue type plasminogen activator in goat milk: generation of transgenic goats and analysis of expression. Biotechnology 9:835–838
27. Baldassarre H, Wang B, Gauthier M et al (2004) Effect of GnRH injection timing in the production of pronuclear-stage zygotes used for DNA microinjection. Zygote 12:257–261
28. Huang YJ, Huang Y, Baldassarre H et al (2007) Recombinant human butyrylcholinesterase from milk of transgenic animals to protect against organophosphate poisoning. Proc Natl Acad Sci U S A 104:13603–13608
29. Zhang J, Li L, Cai Y et al (2008) Expression of active recombinant human lactoferrin in the milk of transgenic goats. Protein Expr Purif 57:127–135
30. Huang S, Zhang K, Huang Y et al (1998) Secretion of biological active human IX protein in the milk of transgenic goats. Chin Sci Bull 43:1317–1319
31. Maga EA, Shoemaker CF, Rowe JD et al (2006) Production and processing of milk from transgenic goats expressing human lysozyme in the mammary gland. J Dairy Sci 89:518–524
32. Ko JH, Lee CS, Kim KH et al (2000) Production of biologically active human granulocyte colony-stimulating factor in the milk of transgenic goat. Transgenic Res 9:215–222
33. Freitas VJF, Serova IA, Andreeva LE et al (2007) Production of transgenic goat (*Capra hircus*) with human granulocyte colony stimulating factor (hG-CSF) gene in Brazil. Anais da Acad Brasileira de Ciencias 79:585–592
34. Freitas VJF, Teixeira DIA, Melo LM et al (2010) Generation of transgenic naturalized goats producing human granulocyte-colony stimulating factor (hG-CSF) in Brazil. Transgenic Res 19:146
35. Lee CS, Lee DS, Fang NZ et al (2006) Integration and expression of goat β-casein/hGH hybrid gene in a transgenic goat. Reprod Dev Biol 30:293–299
36. Lavine G (2009) FDA approves first biological product derived from transgenic animal. Am J Health Syst Pharm 66:518

37. Yekta AA, Dalman A, Eftekhari-Yazdi P et al (2013) Production of transgenic goats expressing human coagulation factor IX in the mammary glands after nuclear transfer using transfected fetal fibroblast cells. Transgenic Res 22:131–142
38. Baruah GL, Nayak A, Winkelman E et al (2006) Purification of monoclonal antibodies derived from transgenic goat milk by ultrafiltration. Biotechnol Bioeng 93:747–754
39. Anema SG, Stanle DJ (1998) Heat-induced, pH-dependent behaviour of protein in caprine milk. Int Dairy J 8:917–923
40. Montilla A, Calvo MM (1997) Goat's milk stability during heat treatment. Effect of pH and phosphates. J Agric Food Chem 45:931–934
41. Ram S, Sindhu JS (1991) Salt balance and heat stability of goat milk as affected by pH. Indian J Anim Sci 61:753–756
42. Zadow JG, Hardham JF, Kocak HR et al (1983) The stability of goat's milk to UHT processing. Aust J Dairy Technol 38:20–23
43. Fonseca CR, Bordin K, Fernandes AM et al (2013) Storage of refrigerated raw goat milk affecting the quality of whole milk powder. J Dairy Sci 96:4716–4724
44. Silanikove N, Leitner G, Merin U et al (2010) Recent advances in exploiting goat's milk: Quality, safety and production aspects. Small Rumin Res 89:110–124
45. Bylund G (2003) Dairy processing handbook. Tetra Pak Processing Systems AB, Lund, pp 361–374
46. Kajal MFI, Wadud A, Islam MN, Sarma PK (2012) Evaluation of some chemical parameters of powder milk available in Mymen singh Town. J Bangladesh Agric Univ 10:95–100
47. Bo L, Chen CL (1990) Effects of formula protein level and ratio on infant growth, plasma amino acids and serum trace elements I. Cow's milk formula. Acta Paediatica 79(3):257–265
48. Janas LM, Picciano MF, Hatch TF (1987) Indices of protein metabolism in term infants fed either human milk or formulas with reduced protein concentration and various whey/casein ratios. J Paediatrics 110(6):838–848
49. Prosser GC, Svetashev VI, Vyssotski MV et al (2010) Composition and distribution of fatty acids in triglycerides from goat infant formulas with milk fat. J Dairy Sci 93(7):2857–2862
50. Lindquist S (2012) Bile salt-stimulated lipase plays an unexpected role in arthritis. Development in rodents. PLoS 7(10):e47006
51. Harvey AJ, Ivarie R (2003) Validating the hen as a bioreactor for the production of exogenous proteins in egg white. Poult Sci 82(6):927–930
52. Lillico SG, Sherman A, McGrew MJ et al (2007) Oviduct-specific expression of two therapeutic proteins in transgenic hens. Proc Natl Acad Sci U S A 104(6):1771–1776
53. Chrenek P, Ryban L, Vetr H et al (2007) Expression of recombinant human factor VIII in milk of several generations of transgenic rabbits. Transgenic Res 16(3):353–361
54. Ling HY, Edwards AM, Gantier MP et al (2012) An interspecific Nicotiana hybrid as a useful and cost-effective platform for production of animal vaccines. PLoS One 7(4):e35688
55. Rodriguez A, Castro FO, Aguilar A et al (1995) Expression of active human erythropoietin in the mammary gland of lactating transgenic mice and rabbits. Biol Res 28(2):141–153
56. Lamming R, Hampson J (1996) The environment as a supply chain management issue. Br J Manage 7:S45–S62
57. Massoud M, Attal J, Thépot D et al (1996) The deleterious effects of human erythropoietin gene driven by the rabbit whey acidic protein gene promoter in transgenic rabbits. Reprod Nutr Dev 36(5):555–563
58. Saunders MJ (1997a) Strategic purchasing and supply chain management. Pitman, London
59. Saunders MJ (1997b) Making strategic decisions and actions in purchasing and supply chain management. In: Proceedings of the 6th international IPSERA conference, pp 61–69
60. David J, Ketchen DJ (2008) Research methodology in strategic management past accomplishments and future challenges. Res Methods 11(4):643–658
61. Chandra C, Kumar S (2000) Supply chain management in theory and practice: a passing fad or a fundamental change? Ind Manag Data Syst 100(3):100–114

62. Olson LD (2012) Supply chain information technology. In: Nahmias S (ed) The supply and operations management collection. Business Expert Press, New York
63. Shaik MN, Abdul-Kader W (2013) Inter organizational information systems adoption in supply chains: a context specific framework. Int J Inform Syst Suppl Chain Manage 6(1):24–40
64. Nelson M (2001) Sustainable competitive advantage from information technology
65. Chandra C, Grabis J (2007) Supply chain configuration – concepts, solutions and applications. Springer, New York
66. Turek B (2013) Information systems in supply chain integration and management. http://www.ehow.com/info_8337099_information-supply-chain-integration-management.html
67. Norris G, Hurley JR, Hartley KM et al (2000) E-business and ERP: transforming the enterprise. Wiley
68. Yanjing J (2009) Integration of ERP and CRM in E-commerce environment. In: Proceedings of the international conference on management and service science, pp 1-9
69. Marakas MG (2003) Decision support systems in the 21st century. New Jersey, Prentice Hall
70. Prem kumar GP (2000) Inter organization systems and supply chain management: an information processing perspective. Inf Syst Manag 17(3):56–69
71. Grover V, Saeed KA (2007) The impact of product, market, and relationship characteristics on inter organizational system integration in manufacturer supplier dyads. J Manag Inf Syst 23 (4):85–216
72. Wisner JD, Tan KC, Leong GK (2012) Supply chain management: a balanced approach, 3rd edn. South-Western Cengage Learning
73. Inderfurth K (1997) Simple optimal replenishment and disposal policies for a product recovery system with lead times. OR Spectr 19(2):111–122
74. Kelle P, Silver EA (1989) Forecasting the returns of reusable containers. J Oper Manag 8 (1):17–35
75. Simpson VP (1978) Optimum solution structure for a repairable inventory problem. Oper Res 26(2):270–281
76. Van der Laan E, Salomon M, Dekker R, Van Wassenhove L (1999) Inventory control in hybrid systems with remanufacturing. Manag Sci 45(5):733–747
77. Wang J, Zhao J, Wang X (2011) Optimum policy in hybrid manufacturing/remanufacturing system. Comput Ind Eng 60(3):411–419
78. Tenunter R, Van der Laan E, Vlachos D (2004) Inventory strategies for systems with fast remanufacturing. J Oper Res Soc 55(5):475–484
79. Diabat A, Govindan K (2011) An analysis of the drivers affecting the implementation of green supply chain management. Resour Conserv Recycl 55(6):659–667
80. Luthra S, Garg D, Haleem A (2014) Green supply chain management. Implementation and performance – a literature review and some issues. J Adv Manag Res 11(1):20–46
81. Beamon BM (1999) Designing the green supply chain. Logist Inf Manag 12(4):332–342
82. Mentzer JT (2001) Supply chain management. Sage Publications
83. Annual Report (2012) European Food Safety Authority – EFSA. https://www.efsa.europa.eu/en/corporate/pub/ar12
84. ICH Guidance for Industry (Q1A) (R2) (2003) Stability testing of new drug substances and products in federal register. ICH Guidance for Industry, Geneva
85. ICH Q1B (1996) Photo stability testing of new drug substances and products. In: International conference on harmonisation of technical requirements for registration of pharmaceuticals for human use. Centre for Biologics Evaluation and Research (CBER)
86. ICH Q3A (2006) Impurities in new drug substances. Center for Biologics Evaluation and Research (CBER). www.ich.org/fileadmin/Public_Web_Site/ICH.../Q3A.../Q3A_R2
87. ICH Q3B (2006) Impurities in new drug products. Centre for Biologics Evaluation and Research (CBER). www.pharma.gally.ch/ich/q3b028295en.pdf
88. ICH Q5C (1995) Quality of biotechnological products: stability testing of biotechnological/biological. Centre for Biologics Evaluation and Research (CBER), Products. www.ich.org/products/.../quality/quality.../stability-testing-of_biotechnologicalbiologi

89. ICH Q6A Specifications (2000) Test procedures and acceptance criteria for new drug substances and new drug products: chemical substances. Center for Biologics Evaluation and Research (CBER). www.ich.org/fileadmin/Public_Web.../ICH_Products/Guidelines/.../Q6A/.../Q6Astep4
90. Arumugam V, Chang HW, Ooi KB et al (2009) Self-assessment of TQM practices: a case analysis. TQM J 21:46–58
91. ICH Q6B Specifications (2005) Test procedures and acceptance criteria for new drug substances and new drug products: biotechnological/biological products. Center for Biologics Evaluation and Research (CBER). www.ich.org/fileadmin/Public_Web_Site/ICH_Products/.../Q6B/.../Q6B_Guideline.pd
92. Mohanty RP, Behera AK (1996) TQM in the service sector. Work Study 45:13–17
93. Prajogo DI, Sohal AS (2003) The relationship between TQM practices, quality performance, and innovation performance: an empirical examination. Int J Qual Reliab Manag 20(8):901–918